NUCLEAR AND RADIATION PHYSICS IN MEDICINE

A CONCEPTUAL INTRODUCTION

Nuclear and Radiation Physics in Medicine

A Conceptual Introduction

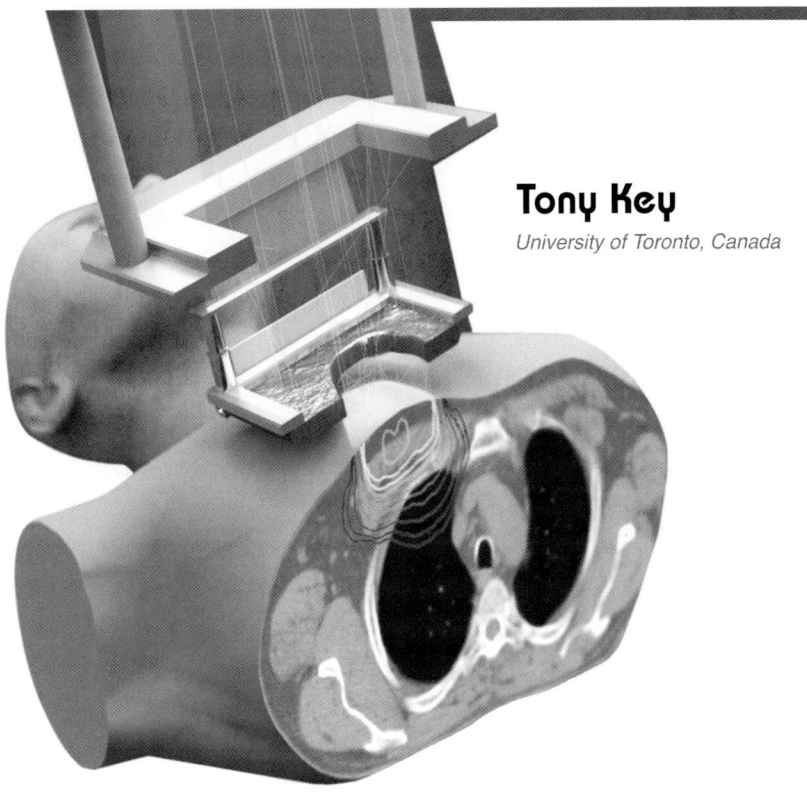

Tony Key

University of Toronto, Canada

World Scientific

NEW JERSEY · LONDON · SINGAPORE · BEIJING · SHANGHAI · HONG KONG · TAIPEI · CHENNAI

Published by

World Scientific Publishing Co. Pte. Ltd.

5 Toh Tuck Link, Singapore 596224

USA office: 27 Warren Street, Suite 401-402, Hackensack, NJ 07601

UK office: 57 Shelton Street, Covent Garden, London WC2H 9HE

Library of Congress Cataloging-in-Publication Data
Key, Tony, 1939–   author.
    Nuclear and radiation physics in medicine : a conceptual introduction / Tony Key.
        p. ; cm.
    Includes bibliographical references and index.
    ISBN 978-9814566803 (hardcover : alk. paper) -- ISBN 9814566802 (hardcover : alk. paper)
    I. Title.
    [DNLM: 1. Health Physics. 2. Nuclear Physics. 3. Radiation. 4. Radiography. 5. X-Rays.
WN 110]
    R898
    616.07'572--dc23
                                                                    2013037381

British Library Cataloguing-in-Publication Data
A catalogue record for this book is available from the British Library.

Cover:

Photo originally appeared as cover of Physics in Canada, Volume 58, No. 2 (2002). The Monte Carlo dose engine was developed by Iwan Kawrakov of NRC and is driven by a new electron beam model developed by MDS-Nordion. The rendering was produced by Tomas Lundberg of MDS-Nordion and was included as Figure 6 in the article by D.W.O Rogers (pp. 63–70) of that issue. Reprinted with permission from the Canadian Association of Physicists.

Printed in Singapore by Mainland Press Pte Ltd.

Preface

This book emerged from a set of notes that I prepared over the years that I taught the final quarter of a large first year physics course at the University of Toronto. Traditionally, the department had offered one high-level course for students intending to specialize, and two courses for the rest. All were 'pure' physics courses, with hardly a nod to applications. Professor Kenneth McNeill, one of the first medical physicists in the department, proposed and designed a course that replaced the non-specialist courses by one course that specifically addressed the life science interests of the large majority of students. The course was hugely successful, accounting for more than 80% of all first year students taking physics. Professor McNeill initially taught *Physics for the Life Sciences* in several sections, each containing about 200 students, at different times during the week. After he retired, the course moved into the only venue on campus that could accommodate the almost thousand students enrolled, and was team-taught in four parts by four different lecturers.

My intention in teaching the final section of the course was to expose students to the fundamentals of radiation and nuclear physics that have had such an important influence on medical practice. I wanted them to know and appreciate the extraordinary intellectual achievements of the scientists whose work is so much a part of our scientific and cultural heritage. I also wanted to show them how some of their newly acquired knowledge of physics could be applied to an understanding of some of the science that underlies medical procedures using X-rays or nuclear physics that they, or those close to them will almost certainly experience at some point in their lives. I also hoped that this brief (twelve lectures) introduction would provide a broad, though certainly not deep, foundation for their further careers in the life sciences.

This book could be used as a text for an introductory course in medical physics or biophysics. Although it represents twelve of my lectures it could easily be adapted to occupy less or more teaching time. Chapter 1 is required for the understanding of the others, but the information therein might well be taught in other sections of a larger course. After chapter 1, each chapter is almost self–sufficient.

I also hope that my work might have a slightly larger appeal. For those who are starting their careers in medical sciences or are already practitioners, it offers some interesting and useful background and an aide-memoire of the basics. For members of the public it could provide a deeper understanding of the science that informs the medical procedures that too many will be subject to, at a deeper level than the often excellent but, of necessity very basic and purely practical information available from hospitals and Web sites. The former audience may be interested in the mathematical demonstrations; the latter certainly will not be. However, for both audiences, the details of the calculations are less important than the knowledge that they can be done.

In the development of the topics I discuss, I assume no knowledge apart from that of the most basic principles of physics: mass, force, momentum, and energy. The medical details are sparse; those that are mentioned are included to provide context and explain the relevance of the physics. Each chapter has a selection of exercises that allow a serious student to test his understanding; a set of worked examples at the end of the book provides models.

Chapter 1 lays the foundation for the rest: the experimental discovery of the atom and its nucleus, the hypothesis that explained the observed atomic spectra, and the mass-energy equivalence. Important units are defined. Chapter 2 focuses on X-rays: their production and use. The main thrust of this chapter is to understand why X-rays are so useful in diagnosis, the parameters that determine the resolution of their images, and the units used to measure their effects on matter. Chapter 3 gives a very brief introduction to radioactivity in nuclei, concluding with a discussion of their use in diagnosis. Chapter 4 introduces the use of radiation in therapy, including X-ray and electron beams and radioisotopes. The units required to make intelligent choices are defined. Chapter 5 is included to give the reader an overview of the immensely complicated area of estimating the health hazards of radiation exposure, with the aim of placing in perspective the radiation risks from medical procedures. Finally, Chapter 6 explains the wonders of Magnetic Resonance Imaging. The ability to see such detail inside the human body would have seemed miraculous in times past. My aim in this chapter is to explain its mysteries without robbing it of its magic.

Acknowledgements

With no training in medical physics, I relied heavily on the work of others. I owe a great debt to Professor Kenneth McNeill and his daughter Diane (McNeill and McNeill, [2002]), who initiated the course for which this text was prepared and who gave me full access and permission to use his exhaustive notes and problem sets. Sadly, the course was discontinued in the year of Professor McNeill's death. Dr Pierre Savaria's more recent notes (P.Savaria, *Notes for PHY238Y* (unpublished, 2002)) and ability at problem solving have also been a great help to me, as have been many suggestions from Dr David Harrison, whose brilliant Flash animations are referenced throughout the early chapters. I much appreciate the support and valuable comments I received from Drs Ruxandra Serbanescu and Mike Bronskill both during the course and the preparation of this book. Drs Trevor Levere, Rashmi Desai, and Yau Chan gave me kind and important support in the early stages of preparation. The many excellent teaching assistants with whom I worked gave valuable feedback to my notes, and solutions for many of the problems. Michiya Sasaki of ICRP was helpful beyond her duty in assisting my search for some of the data used in Chapter 4. Without Raul Cuhna's extraordinary knowledge and talent, and his good nature and unflappability under stress, the figures in this book would be unrecognizable. This book would not have been possible without the support of my colleagues, both faculty and staff, in the Department of Physics over my many years of teaching. Finally, I want to thank my wife Mervyn, whose patience and support was unwavering in the face of my labours.

Toronto,
June 2013

Contents

Chapter 3 Radioactivity and Radioisotopes

Chapter 4 Radiation Therapy

List of Tables

List of Figures

Chapter 1

Introductory Atomic and Nuclear Physics

1.1 Introduction

We have become so accustomed to the idea that matter is made of tiny parts— molecules, atoms, and electrons—that it is somewhat of a surprise to realize that this fact became accepted just over a hundred years ago. Indeed it is rumoured that Ludwig Boltzmann, depressed that the rest of the scientific world did not share his belief in the reality of molecules, committed suicide in 1906. Boltzmann had shown that the results of classical thermodynamics could be beautifully explained by the assumption that matter consists of small, indivisible, weakly interacting molecules. The equation that appears on

Figure 1.1. Ludwig Boltzmann [Courtesy AIP Emilio Segre Visual Archives.]

his gravestone — $S = k \ln W$ — summarizes his work; it relates the macroscopic entropy (S) of a system to the number of microscopic states (W) that the system can access, where k is Boltzmann's constant.

Of course, Boltzmann acted precipitously, since there were many supporters of the theory that matter consisted of indivisible atoms or molecules: Democritus in Greek times, Newton in the 17th century, Dalton and Avogadro in the 19th. Although some scientists continued to deny reality to objects that could not be observed directly, experimental results were against them; atoms and molecules are much more than merely mathematical fictions. Even more astonishingly, experiments showed that the atom was not indivisible; atoms, apparently, consist of other, even smaller objects, whose reality is also incontrovertible.

1.2   The discovery of the nucleus

In 1897, J.J. Thomson of Cambridge University was studying the passage of electric current through dilute gases contained in glass tubes called discharge tubes. He noticed that particles were emitted from the negative poles—the cathodes—of the tubes. These particles were shown to have a negative charge; we now call them electrons. Since they are emitted from atoms, it seems probable that they are constituents of the atom. Since the atom is neutral, the rest of the atom has to have a positive charge. Accordingly, Thomson proposed the 'Plum Pudding' model, in which the electrons are embedded in a uniform sphere of positive charge, like the plums in a pudding. That this model was a poor one was demonstrated by the brilliant experiments of Ernest Rutherford and his collaborators.

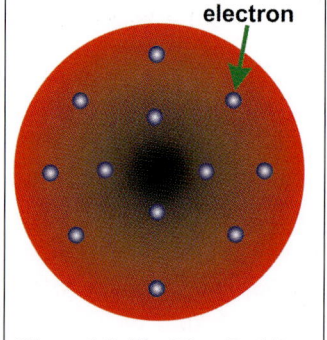

Figure 1.2. The Plum Pudding model.

In 1896, the French physicist Antoine Henri Becquerel, noticed that some rocks that he had been studying emitted radiation that could pass through opaque photographic paper. Ernest Rutherford ("the father of Nuclear Physics"), a New Zealander, and one of Thomson's students, identified two of these radiations. He showed that the first, which he called **alpha** radiation, is nothing but Helium atoms with two electrons removed, with, therefore, a positive charge of plus two. His experiments using these alpha particles showed conclusively that the mass of the atoms is almost totally concentrated in a tiny 'nucleus' at the centre.

The experiments were simple in principle: a beam of alpha particles from a radioactive source were directed at a thin gold leaf. If the Plum

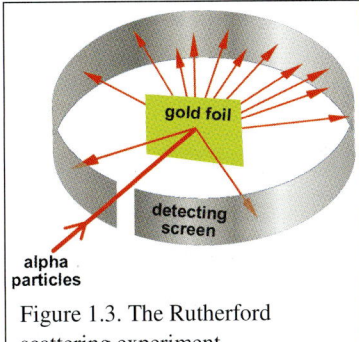

Figure 1.3. The Rutherford scattering experiment.

Pudding model represented reality, the alpha particles would be expected to suffer only many small deflections as they encounter the spread-out charge of the gold atoms. However, Rutherford, much to his astonishment, noted a significant number of very large deflections, some of them even causing the heavy alpha particles to bounce back from the foil.

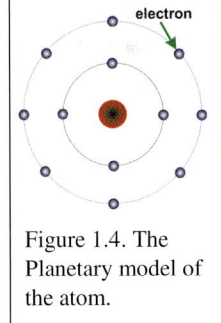

Thus was born the Planetary Model of the atom, in which the electrons circulate around a tiny but very massive nucleus, rather as the planets circulate round the sun.

Figure 1.4. The Planetary model of the atom.

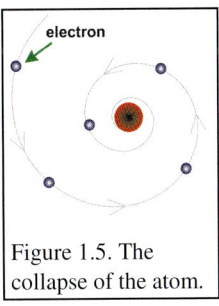

Figure 1.5. The collapse of the atom.

Unfortunately, Rutherford's planetary model has an apparently fatal flaw. The accelerating electrons that were supposed to orbit around a central nucleus would, according to classical physics, lose energy by emitting electromagnetic radiation (see Appendix A1.1); this loss of energy would cause them to spiral inward to the nucleus till the atom collapsed. That we are here to observe our world lets us know that this does not happen—so what's going on here?

There was another difficulty. Atoms do indeed emit electromagnetic radiation. However, rather than a spectrum that has a continuous range of energies, as classical physics predicts, this radiation appears only at a finite number of discrete energies, shown in Figure 1.6. Both difficulties were 'solved' by Niels Bohr, the Danish genius (§1.5).

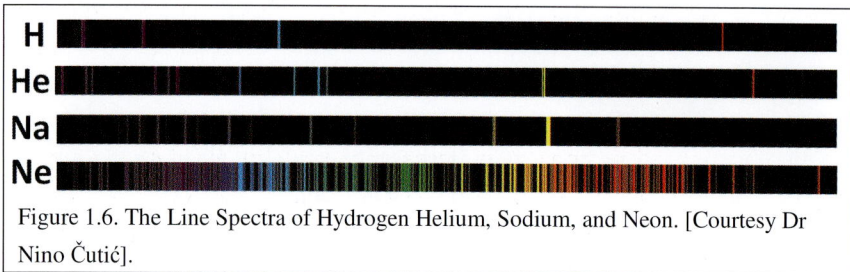

Figure 1.6. The Line Spectra of Hydrogen Helium, Sodium, and Neon. [Courtesy Dr Nino Čutić].

1.3 The basics of nuclear physics

We now know that the nuclei of atoms consist of **protons** with positive charge, equal and opposite to that of the electron, and **neutrons** with charge zero; together these are called **nucleons**. The number of protons in a nucleus is denoted by Z and the number of neutrons by N. Z is called the **atomic number** of the nucleus since it indicates where on the table of the elements that particular one lies. The **mass number** of the nucleus—the total number of nucleons—is denoted by A, which equals $N+Z$. Then a particular nucleus **X** is written as $^{A}_{Z}X$—e.g. $^{1}_{1}H$, $^{12}_{6}C$, $^{16}_{8}O$, etc. Since Z is entirely determined by the chemical symbol, this is often abbreviated by omitting the Z value, as in ^{12}C, ^{16}O: pronounced "Carbon–12, Oxygen–16". Nuclei that have the same value of A are called **isobars**.

The value of Z determines the chemical behaviour of the element. However, different nuclei can have the same value of Z, yet different values of N (and thus of A). Nuclei with the same Z but different A values are called **isotopes**. Thus ^{12}C, ^{11}C, and ^{14}C are isotopes of Carbon. Isotopes are either stable or unstable; the unstable ones emit radiation, as we shall see in Chapter 3, and are called **radioisotopes**.

Rutherford's experiments indicated the approximate size of the gold nuclei he used as targets. Later experiments showed that the nuclei of all atoms are roughly spherical with a radius that depends on the number of nucleons. This radius is given by $R = r_0 A^{1/3}$ where $r_0 \approx 1.2 \times 10^{-15}$ m. (10^{-15} m is defined as one fermi after the famous Italian physicist; it is also one femtometre; the abbreviation is fm). This is an astonishingly small number in comparison to the radius of the atom of about 10^{-10} m.

Since nuclei hold together, in spite of the Coulomb repulsion between the positively charged protons, it is necessary to postulate another force of nature—the so-called strong force—that comes into play only at a distance of less than 10^{-15} m or so. This very short-range strong force does not distinguish between neutrons and protons, being attractive between any two nucleons (p–p, p–n, n–n); it is the glue that holds the nucleus together. For very large nuclei, the peripheral protons can lie outside of the range of the strong force and their mutual Coulomb repulsion tends to pull the nucleus apart. However the strong attractive forces between the neutrons tend to moderate this effect. The interplay of these opposing forces sets a limit on the size of stable nuclei.

Just as we would need to apply energy to raise a stone out of a hole in the earth, energy is needed to pull a nucleus apart. Since Albert Einstein showed that, energy is mass (§1.4.2), this means that the sum of the masses of the individual protons and neutrons of which a nucleus is composed is greater than the mass of the nucleus. The difference between the sum of the masses of the nucleons and the mass of the nucleus that they make up, expressed as energy, is called the **binding energy** of the nucleus.

1.4 Notes on units

1.4.1 *The electron volt*

The SI unit of energy, the joule, is too large a unit to be useful in atomic or nuclear physics. Accordingly we define the **electron volt (eV)**, the energy that an electron would acquire when accelerated through a potential difference of one volt. The electron is now known to have a charge of $e = 1.60 \times 10^{-19}$ Coulombs (first measured by Robert Andrews Millikan's famous oil-drop experiment in 1909). Thus the electron volt is related to the joule by 1 eV $= 1.60 \times 10^{-19}$ joules.

1.4.2 *The most famous equation in the world*

For the very small masses encountered in atomic or nuclear physics, the kilogram is not a useful unit (the mass of the electron, for example, is 9.11×10^{-31} kg). A more useful unit is defined using the most famous equation in physics, $E = mc^2$, a result of Einstein's theory of Special Relativity (1905). E is the energy, m is the mass, and $c \cong 3 \times 10^8$ m.s^{-1} is the velocity of light. This equation quantifies the fact that mass and energy are equivalent, so that mass can be expressed in energy terms. For example, if we could somehow completely annihilate an electron, Einstein's equation shows that the energy would be:

$$E = (9.11 \times 10^{-31} \text{ kg}) \times (3.00 \times 10^8 \text{ m/s}^{-1})^2$$
$$= (8.20 \times 10^{-14} \text{J})/(1.60 \times 10^{-19} \text{J/eV}) = 0.511 \text{ MeV} \qquad (1.1)$$

Thus we can write the electron mass ($m_e = E/c^2$) as 0.511 MeV/c^2. **MeV/c^2** is a useful unit, since it gives at a glance how much energy a particular mass could yield.

1.4.3 *The atomic mass unit*

Another mass unit used in nuclear physics is the **atomic mass unit** (**amu**, abbreviation **u**), defined as the mass of 1/12 th of the mass of the ^{12}C atom*. ^{12}C is chosen since it is abundant and extremely stable. Its mass has been measured to be $1.992\ 671 \times 10^{-26}$ kg. Thus 1u = 1.660 539 $\times 10^{-27}$ kg. In these units the masses of the electron, proton, and neutron are $m_e = 5.485\ 799 \times 10^{-4}$ u, $m_p = 1.007\ 276$ u and $m_n =$

---

* This is more properly called the **unified mass unit**. The atomic mass unit was originally based on ^{16}O. However, many authors continue to use the term 'atomic mass unit' and its abbreviation amu for the newer unit based on ^{12}C, a practice that is followed in this book.

1.008 665 u . Since $c = 2.977\,924 \times 10^8 \text{m.s}^{-1}$, a conversion factor to MeV/c^2 can be derived by calculating the energy contained in a mass of 1u $\{E(u)\}$:

$$
\begin{aligned}
E(u) &= (1.660\,539 \times 10^{-27} \text{kg}) \times (2.997\,924 \times 10^8 \text{ m/s})^2 \text{ J} \\
&= (1.492\,435 \times 10^{-10} \text{J})/(1.602\,177 \times 10^{-19} \text{J}/eV) \\
&= 931.49 \text{ MeV}
\end{aligned}
\tag{1.2}
$$

Thus 1 atomic mass unit has a mass of 931.49 MeV/c^2.

1.4.4 *The mole and Avogadro's number*

Here is the modern definition of a mole that we will use. **1 mole of an element is defined as that quantity of the element that contains exactly as many atoms as there are in 12 grams of ^{12}C.** This number is called **Avogadro's number,** N_A; it equals 6.022×10^{23} mole^{-1}. Thus one mole of ^{12}C contains exactly N_A atoms and has a mass of EXACTLY 12g, by definition . This **molar mass** of 12g is just the mass of each ^{12}C atom in amu multiplied by Avogadro's number, i.e. $12u \times N_A = 12\text{g}$; so we see that $1u = (1/N_A)\text{g}$. In general the molar mass of an atom is given by the mass of the atom multiplied by N_A. Thus the molar mass of a nucleus ^{A}X with mass number A is very nearly, but, due to the nuclear binding energy, **not exactly**, equal to A grams.

1.5 The Bohr model of the atom

To address the difficulties inherent in the planetary model of the atom mentioned in §1.2, Bohr arbitrarily stated two postulates.

 i. Electrons in atoms do not obey classical physics; instead they exist in 'stationary' states (i.e. states that do not radiate electromagnetic energy by definition). The possible energies of these states are discrete, not continuous—i.e. they can occur only with specific values

 ii. Atoms do emit electromagnetic radiation; however they do so only by jumping from states of higher energy to states of lower energy. The energy of the emitted radiation is thus equal to the difference in energies between these states.

The first postulate, although it makes no sense from the standpoint of classical physics, allowed the further development of the atomic model.

The second postulate explains the observation of discrete atomic spectra, shown in Figure 1.6. Since there are only specific energy states available to electrons in an atom, the energies of the emitted radiation, acquired by jumps between these states, must also have only specific values. To calculate the values of these energies, Bohr used quasi-classical ideas that gave a surprisingly good explanation of the experimentally observed spectral lines of Hydrogen.

The energy level diagram for Hydrogen is shown in Figure 1.7; the names of the experimenters who first studied the different series are indicated. (The four lines of the visible Balmer series can be seen in the H spectrum shown in Figure 1.6). The possible states are labeled by the 'principal quantum number', n, where n is an integer taking all positive values from one to infinity. The energy of each state is then given by $(- C/n^2)$, where C is a constant for the atom under consideration (equal to 13.61 eV for Hydrogen). The minus sign takes account of the fact that in order to remove an electron from an atom, positive energy must be applied (think again of lifting a stone from the bottom of a well).

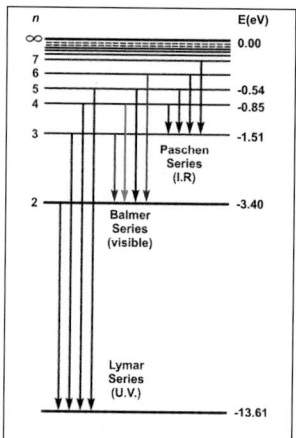

Figure 1.7. The energy levels of the Hydrogen atom.

For instance if exactly $+13.61/3^2$ eV were applied to an electron in the third energy level of Hydrogen, the electron would just escape from the atom, with no remaining kinetic energy. If less than this amount of energy were applied, the electron could not escape; if more than this amount were applied, the electron could escape with positive kinetic energy (equal to the difference between the energy applied and $+13.6/3^2$ eV).

The energy of the radiation emitted by a transition from the nth to the mth energy levels is given by $\{- C (1/n^2 - 1/m^2)\}$. The minimum energy required to remove an electron from an atom is the **ionization energy**; the ionization energy of Hydrogen is thus $+13.6$ eV.

There were many more or less successful attempts to extend Bohr's simple picture to multi-electron atoms, but a consistent and accurate theory had to await the development of quantum mechanics and the discovery of electron 'spin' (discussed in Chapter 6). The electrons in multi-electron atoms lie in shells, each shell containing a few levels of closely similar energies that depend in a complicated way on the electron spins. Each shell corresponds to the 'principal quantum number', n, that can take the values 1, 2, 3, ... It turns out that nuclei also have discrete energy levels, as we shall see in Chapter 3.

1.6   The photon

Whether light is particle-like or wave-like has been a topic of controversy for over 400 years since Isaac Newton, Thomas Hooke, and Christian Huygens, among others, debated it in the 17th century. Thomas Young's famous double slit experiment in 1801 seemed to settle the question; light is undoubtedly wavelike in nature. Its colour is determined by its wavelength, λ, which is related to its frequency, ν, by the equation $c = \lambda\nu$, where c is the speed of light in vacuum. However, in 1905, Einstein's explanation of the **photoelectric effect** demonstrated, equally conclusively, that a ray of light consists of a stream of particles that we now call photons!

Finally, quantum mechanics resolved the paradox by accepting it. Nothing in our everyday experience allows us to form a visualization of an object that sometimes looks like a wave (which spreads out, and exhibits interference), and sometimes like a particle (which is located at a point in space, and does not exhibit interference). That concept, and many others that apparently fly in the face of common sense, is one we must get used to if we want to understand the world of the very small, where quantum mechanics holds sway.

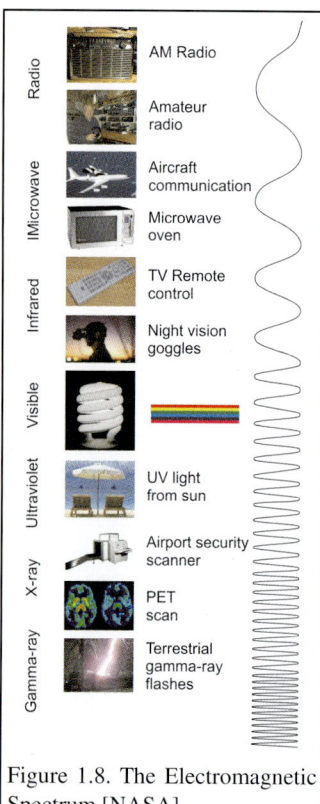

Figure 1.8. The Electromagnetic Spectrum [NASA].

Electromagnetic waves cover a huge range of frequencies, from radio waves (first discovered by Heinrich Hertz) to X-rays and gamma rays; visible light occupies only a tiny portion of the spectrum (Figure 1.8).

For most of our discussion of medical applications, it will be most useful to consider the electromagnetic radiation that will interest us—X-rays or gamma rays—to be particles, each with an energy E that is related to the frequency of the wave, v, by $E = hv$, where h is called **Planck's constant**; as with all waves, the speed of the wave, c, is given by $c = \lambda v$ where λ is the wavelength (see Appendix A1.1). You will notice the 'schizophrenic' nature of the last equation, which has a particle property (energy) related to a wavelike property (frequency). Such is the world we inhabit.

A1.1 *Electromagnetic radiation*

In a series of brilliant experiments spanning the years 1831 to 1839, the English scientist Michael Faraday demonstrated that a changing magnetic field causes electrons to move in a conducting medium. We interpret this to mean that a changing magnetic field generates an electric field that causes the electrons to move. But moving electrons constitute an electric current, and, as the Danish physicist, Hans Christian Ørsted, showed in 1820, an electric current generates a magnetic field. (During a lecture, Ørsted noticed the deflection of a compass needle when an electric current was passed through a neighbouring wire). Since the electrons have their own electric field, a current of electrons produces a moving electric field. To summarize: a uniformly moving magnetic field generates a constant electric field, and a uniformly moving electric field generates a constant magnetic field.

Now consider what happens when we generate an *oscillating* electric field, for instance by waving an electric charge in the air. The generated magnetic field also oscillates; and this oscillating magnetic field will, in turn, generate another oscillating electric field, with its own associated oscillating magnetic field, and so on! What happens to this apparently self-generating system? The electric and magnetic fields, inextricably joined,

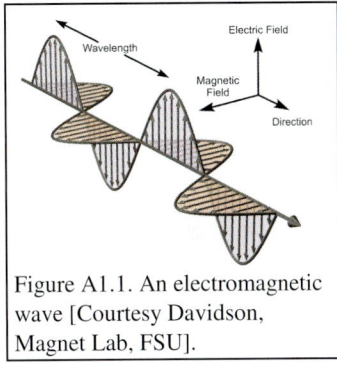

Figure A1.1. An electromagnetic wave [Courtesy Davidson, Magnet Lab, FSU].

oscillating at right angles to each other, dissociate from the accelerating charge and speed off—at the speed of light—as *electromagnetic waves*.

Here is a condensed (and inaccurate!) mathematical sketch to make this process plausible.

Faraday's Law states that an electric field (E), is produced by a changing magnetic field (B): $\quad E \infty \, dB/dt$ $\hspace{2cm}$ (A1.1)

Ørsted's observation shows the magnetic field (B) is proportional to the current that produces it; the current (I) is simply the flow of electrons that carry their own electric field (E) .

Thus: $\hspace{3cm} B \infty I \infty \, dE/dt$ $\hspace{2.5cm}$ (A1.2)

Combining A1.1 and A1.2 we obtain:

$$d^2 E/dt^2 = d(dE/dt) \infty \, dB/dt \infty E \hspace{1.5cm} (A1.3)$$

Similarly: $\; d^2 B/dt^2 = d(dB/dt) \infty \, dE/dt \infty B \hspace{1.5cm}$ (A1.4)

The solution of these equations yields oscillatory behaviour (sines and cosines) for E and B. (Remember that the second differential coefficient of a sine or a cosine returns the sine or the cosine respectively; e.g. $d^2(\sin x)/dt^2 = -\sin x$). The oscillating E and B fields turn out to be at right angles to each other and to the direction of the wave.

The correct mathematical treatment leads to four equations, called Maxwell's equations after the Scotsman James Clerk Maxwell. They brilliantly summarize all of electricity and magnetism. Maxwell showed that the speed of these waves in vacuum was given by a very simple formula which includes only the permittivity and permeability of free space. This is the speed of light in vacuum, $c \cong 3 \times 10^8 \text{m.s}^{-1}$.

Maxwell's work seemed to confirm the belief that light is a wave, and demonstrated that it is the E and B fields that 'waved'. In 1905, Albert Einstein had a 'magical' year, producing no less than three papers, each one of which changed our understanding of the world. The most well-known of these papers (*'On the Electromagnetic Dynamics of Moving Bodies'*) addressed some of the difficulties that arose from this new understanding of the nature of electromagnetic radiation; thus was born the theory of Special Relativity. The second paper began to suggest that this classical view of electromagnetic waves was not the whole story (see Appendix A1.2).

A1.2 *The quantum world*

Many books have been written about the development of quantum mechanics, and controversy around its extraordinary results still appears in the scientific journals. The following ridiculously brief summary is the least that a well-educated person should know.

When any material is heated, it emits light; the visible part of the emitted spectrum is red at lower temperatures, becoming yellow and then white as the temperature increases. Given the knowledge about the origin of electromagnetic radiation in the oscillating charges of the heated material, the theoretical calculation of the observed spectra should have been easy. In fact, calculations based on classically correct assumptions failed spectacularly. In 1900 Max Planck realized success by assuming that the light emitted from the oscillating charged particles, is emitted in lumps, or packets, rather than in the continuous manner that Maxwell's equations predict. The energy of these lumps is proportional to the frequency of the light, and the constant of proportionality was, appropriately, called **Planck's constant**, denoted by h. Initially, this seemed to be an intriguing mathematical trick, with no real physical significance.

However, five years later, the second of the Einstein's 1905 papers showed that Maxwell's 'classical' view of electromagnetic waves could not explain some of the details of the phenomenon of the photoelectric effect (to be discussed briefly in §2.3.1). This effect—in which light knocks electrons out of a metal—can be explained only if light is also made of lumps, which we now call photons. The energy (E) of these photons has to be related to the frequency of the light (v) by the formula $E = hv$, where h is Planck's constant. Planck received the Nobel Prize for his work in 1918; Einstein, for his, in 1921.

Astonishingly, it turns out that particles (electrons, protons, neutrons, etc.), under certain circumstances, behave as waves! This extraordinary feature of nature was first proposed by de Broglie in 1923, and experimentally observed by another Scotsman, G.P. Thomson, among others. The latter was the son of J.J. Thomson, which caused a wag to remark that J.J. proved that the electron was a particle and his son proved that it was a wave! However further discussion will take us too far afield.

Exercises Chapter 1

1. Two protons are separated by a distance of A) 10^{-10} m, and B) 10^{-15} m. Which of the electric force or the strong force is the strongest in these two cases? Provide evidence for your decision in each case.
2. Approximately how many nucleons do you have in your body? State your assumptions clearly.

3. Calculate A) the binding energy and B) the binding energy per nucleon of ^{197}Au. C) What is the exact value of the molar mass of ^{197}Au? D) If gold costs around \$800 per ounce, how much does an atom of gold cost? (You will need to consult tables of atomic masses.)

4. Rutherford's alpha scattering experiment used a gold foil (^{197}Au) of thickness 0.6 μm., density 19.3 g cm^{-3}. Consider alpha particles incident on a gold foil as shown in Figure 1.3. A) Approximately how many atomic layers do the alpha particles encounter as they pass through the foil? State any assumptions you make about the arrangement of the gold atoms in the foil. B) Approximately what is the fraction of the total area of the first layer of the foil that is occupied by nuclei as 'seen' by the beam of alpha particles?

5. If your head represents the size of the nucleus, how far west do you have to go to encounter the electron in the first Bohr orbit? (You may need to consult a map!).

6. In free space, the neutron can decay into a proton and an electron and another particle, which Fermi named a neutrino ('the little neutral one'). How much energy is available for this decay? Why can a proton in free space not decay by a similar mechanism into a neutron and an electron and a neutrino?

7. Line spectra observed when a current is passed through hydrogen are described by the formula $1/\lambda = R(1/n^2 - 1/m^2)$ where R is the Rydberg constant, and n and m are integers. A) Starting from the equivalent formula for these line spectra given in §1.5, calculate the value of R when λ is in m. B) Do the numbers shown in Figure 1.7 correspond to this claim that the energy of a given level of Hydrogen is proportional to $1/n^2$? Show your proof!

8. A hydrogen atom in its ground state absorbs a photon of energy 12.094 eV. To which energy level is the atom excited?

9. Calculate the wavelength of photons of energy 50 keV. What kind of electromagnetic radiation are they? (Check Figure 1.8)

10. A high energy X-ray machine produces 10^{20} X-ray photons per second, each of energy 5 MeV. If these are confined to a beam of cross section 2.5 cm^2, what is the energy intensity (the energy fluence, §4.3.1) of this radiation in W m^{-2}?

11. ^{235}U is used to generate nuclear power—and atomic bombs. When a 'thermal' neutron (a very slow neutron, whose speed is about that of a molecule in a gas at room temperature) interacts with a nucleus of ^{235}U, the ^{235}U fissions, breaking up into several other nuclei and two or more thermal neutrons. These neutrons can then interact with other ^{235}U nuclei, which in turn produce more neutrons, thus generating a 'chain reaction'. Controlled chain reactions are used to generate power; uncontrolled chain reactions energize the atomic bomb. One of the main ways the ^{235}U fissions is via:

n + ^{235}U → ^{236}U → ^{144}Ba + ^{89}Kr + 3n. Since the natural occurrence of ^{235}U is so low, nuclear power stations often use enriched uranium, which contains about 3% to 4% of ^{235}U. (The main technical difficulty faced by the Los Alamos scientists who were designing the atomic bomb that was dropped on Japan in 1945, was separating enough of the rarer ^{235}U). The Pickering Nuclear Generating Station on the shores of Lake Ontario is one of the largest in the world. At peak power, it can generate about 4,000 MW. Approximately what mass of ^{235}U must be converted to energy to allow the Pickering Generating Station to operate at full power for 24 hours? Assume that all of the power produced at Pickering comes from the reaction above (not true, but the other possible reactions produce similar amounts of energy)

Chapter 2

X-rays—Production, Characteristics, and Use

2.1 Introduction

In 1895, Wilhelm Roentgen discovered a mysterious penetrating radiation emitted when beams of energetic electrons, traveling in an evacuated glass tube, hit the glass wall. (He was awarded the first Nobel prize in 1901 for this work). We now know that these rays, which Roentgen named X-rays, are electromagnetic waves that have frequencies in the range of about 10^{18} to 10^{21} Hz, corresponding to photon energies in the range of a few keV to several MeV. Radioactive nuclei also emit electromagnetic waves called gamma rays that have higher energies in the MeV range (discussed in §3.2.4). Figure 1.8 gives a picture of the entire electromagnetic spectrum. This section discusses the physics of the production of X-rays, the interaction of X- and gamma rays with matter, and the application of X-rays to medical diagnosis.

2.2 Production of X-rays

X-rays are produced when high energy electrons strike a metal target. There are two distinct mechanisms of production.

2.2.1 *Bremsstrahlung radiation*

An accelerated charge emits electromagnetic waves (Appendix A1.1). Electrons, shot into a metal target, occasionally pass close enough to a nucleus to be accelerated by the Coulomb force of the positive nucleus,

16

which results in the emission of X-ray radiation. In any one interaction, the electron can lose some or all of its kinetic energy, which is carried off by an X-ray photon. The energy of the X-rays emitted can vary continuously from a maximum (corresponding to the case when the incoming electron loses all of its energy) to very low values. Einstein's work on the photoelectric effect (A1.1) showed that energy of the photon (E_γ) is given by $E_\gamma = h\nu = hc/\lambda$ where ν is the photon frequency, λ its frequency, and h is Planck's constant. Thus the 'cut-off' at the highest energy the emitted X-rays can reach, lead to a *high frequency* cut-off or, equivalently, a *low wavelength* cut-off, followed by a continuous range of wavelengths at higher values of wavelength. The probability of bremsstrahlung radiation increases as the square of the energy of the electron and linearly as the atomic number of the target.

2.2.2 Characteristic X-rays

If the electrons entering the metal target have high enough energy, they may have, in addition to collisions mediated by the Coulomb interaction, more direct collisions in which an electron from one of the lower energy levels of the atom is removed. Atomic electrons from higher energy levels of the atom then cascade down to fill the vacancy in the shell, emitting X-radiation that is characteristic of the atoms of the target. The energy of the radiation is given by the difference between the initial and final atomic energy levels: $E_\gamma = h\nu = E_i - E_f$.

Electrons in atoms lie in shells (§1.5, and Figure 1.7), each shell containing a few levels of closely similar energies. Each shell corresponds to the 'principal quantum number', n, that takes values 1, 2, 3 .... The shell with n =1 is called the K shell, that with n = 2 the L shell, that with n=3 the M shell, etc. The characteristic X-rays are labeled by the shells where the vacancy occurred; often different Greek suffixes (α, β, etc) are added to identify the shell from which the electron cascades down (α from the shell one above, β from the shell two above, etc.).

A schematic of the X-ray spectra from a tube with a molybdenum target is shown opposite for a variety of tube voltages. The smooth curves from the bremsstrahlung radiation display the low wavelength cut-offs, whose values depend on the tube voltage; superimposed on the spectrum are two spikes appearing at the highest tube voltage of 25kV, corresponding to the X-rays that are characteristic of molybdenum.

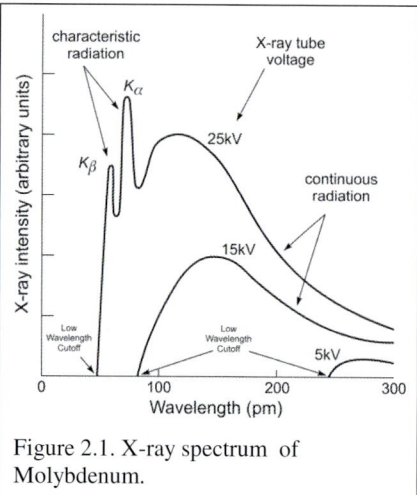

Figure 2.1. X-ray spectrum of Molybdenum.

2.2.3 X-ray machines

As the electrons are decelerated by multiple collisions with the atoms and nuclei of the target, only a small fraction of their energy (typically less than 1%) is given up as X-rays. Generally, the higher the energy of the electrons, the greater is the fraction of their energy that is converted to X-ray photons. The remainder causes heat generation in the target, and steps must be taken to dissipate this heat.

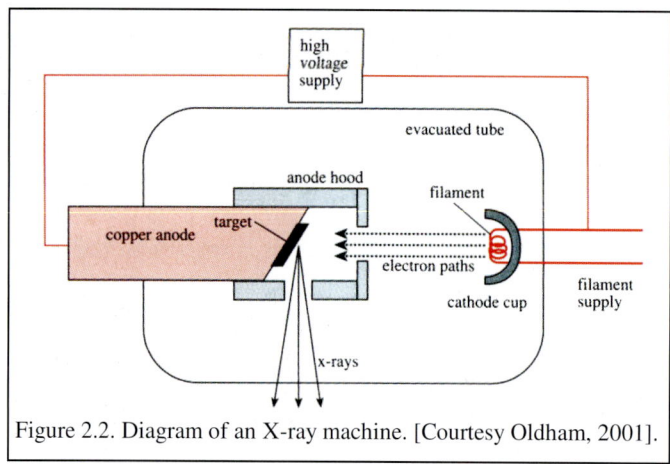

Figure 2.2. Diagram of an X-ray machine. [Courtesy Oldham, 2001].

A diagram of an X-ray tube is shown in Figure 2.2. Electrons emitted from a heated filament are accelerated by a high voltage (typically ~10 to ~100 kV). They strike a target, made of a heavy metal—often lead, molybdenum or tungsten—whose high-Z nuclei maximize the electron's acceleration in the Coulomb field. The target is embedded in copper to provide a good conducting path for heat dissipation.

If the manufacturer quotes a single value for the energy of an X-ray machine, the assumption being made is that the full spectrum of X-rays has the same effect as a mono-energetic beam that has **one-third** the value of the maximum energy of the actual spectrum. We will use this approximation to simplify some calculations.

The low energy X-rays from the tube shown above, typically around several tens of keV, are ideal for diagnostic purposes. However, therapeutic X-rays need to have much higher energies (from around 100 keV for treating superficial skin tumours, for example, up to 25 MeV or so for deeply situated pelvic tumours). For X-rays at these high energies, a linear accelerator is used to accelerate the electrons. A schematic view of a high-energy X-ray machine and a photograph of it in a clinical setting are shown below.

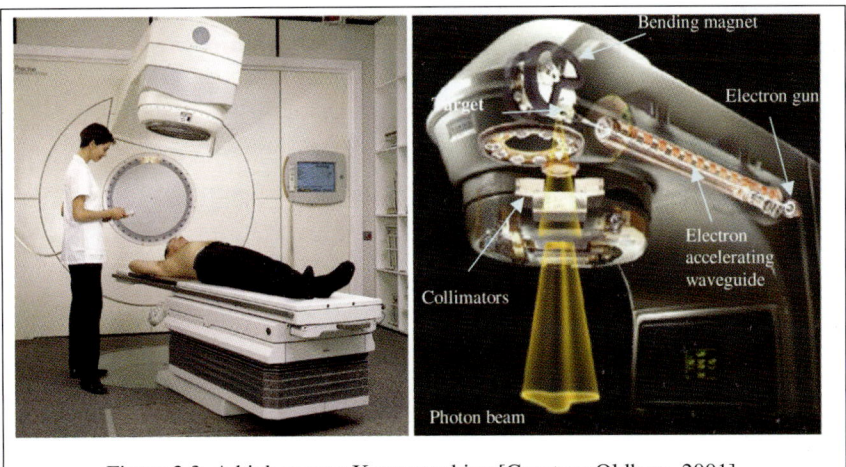

Figure 2.3. A high energy X-ray machine [Courtesy Oldham, 2001].

2.3 The interaction of photons with matter

When a beam of X-rays or gamma rays passes through matter, it is attenuated (i.e. photons are lost to the beam) by interactions with the atoms and electrons of the matter. The amount of attenuation depends on the energy of the rays and the composition of the matter. The differences in attenuation shown by different materials give the X-rays their ability to discriminate between bone, muscle, and tissue. Some molecular damage is also done, which will be our concern when we consider the dangers of radiation. Generally, X- and gamma rays produce electrons that, in turn, lose energy by collisions with atomic electrons or nuclei via the Coulomb interaction. Photons have three main methods of producing electrons.

2.3.1 *The Photoelectric effect*

For photon energies of less than a few tens of keV or so, their main interaction with matter is via the photoelectric effect. Here the photon is totally absorbed by the atom, giving up all of its energy. The excited atom then de-excites by emitting an electron. (As mentioned in §1.6, Einstein's explanation of this effect demonstrated the particle-like nature of light). If the photon has an energy $e_\gamma = h\nu$, and the binding energy of the electron to the atom is B_e (the energy required to remove the electron from the atom) then the energy of the ejected electron is $e_e = e_\gamma - B_e$, (since $B_e \sim$ a few eV, we can usually set $B_e = 0$ for the X-ray energies that interest us).

The probability of an X-ray undergoing the photoelectric effect per atom decreases with increasing X-ray energy since if the energy of the photon is too high it will knock the electron it strikes right out of the atom (the interaction is then classified as the Compton effect, see below). The probability of the photoelectric effect increases quickly with the atomic number of the matter being irradiated (see Equation A2.7). This is because a *free* electron cannot absorb a photon and simultaneously conserve energy and momentum. The photon momentum in the photoelectric effect must be absorbed by the whole atom, and larger atoms with higher Z values have more tightly bound electrons close to the nucleus available to perform this function.

2.3.2 *The Compton effect*

For energies greater than a few tens of keV or so, the effect first explained by Arthur Compton begins to dominate the interaction of photons with matter. Here the photon knocks an electron out of the atom and continues with lower energy on a deflected path; energy and momentum can be simultaneously conserved, and the collision can be considered to be between the photon and a free electron, the electron's binding energy being negligible compared to these X-ray energies. The deflected photons can still be in the X-ray range, so shielding must be provided near X-ray machines. The probability of Compton scattering per atom increases with energy from low values (where the photoelectric effect dominates) and then decreases approximately linearly. It increases with the number of electrons available per atom of the scattering material, and is thus approximately proportional to Z (Equation A2.9).

2.3.3 *Pair production*

In pair production, the equivalence of energy and mass (§1.4.2) is manifested; at high energies, the photon energy converts directly into mass when it passes close to a nucleus, producing an electron-positron pair. The positron, the *anti-particle* of the electron identified by the US physicist Carl Anderson in 1932, is a particle that has exactly the same mass as the electron but opposite charge. The energy of the photon has to be at least the sum of the masses of the pair (1.02 MeV/c^2) for the process to be energetically possible. In practice, the probability of pair production is important in X-ray scattering only above several MeV or so, after which it increases with E and Z.

2.3.4 *Summary*

Figure 2.4 shows approximately how the different processes behave as a function of X-ray energy in biological matter. Appendix A2.2 gives a more complete discussion of the mathematical derivation of these probabilities.

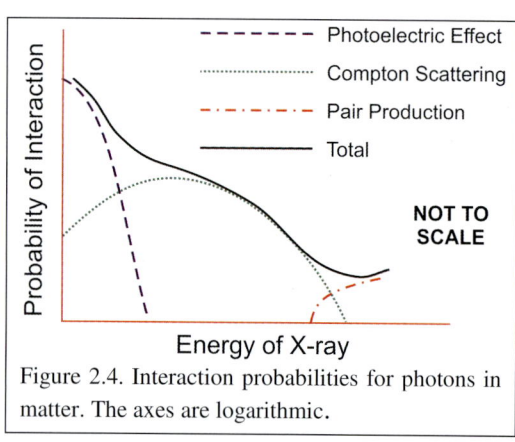

Figure 2.4. Interaction probabilities for photons in matter. The axes are logarithmic.

These interactions change the direction or the energy of the original X-ray or gamma ray from the beam; we use the word **attenuation** to describe this process. However the photon energy is not lost, but is transferred to the secondary electrons, photons, and positrons. If these secondaries interact with the atoms of the material being irradiated, thus depositing some or all of their energy in the irradiated matter, **absorption** of the beam is said to have occurred. The electrons, being light, undergo many small Coulomb scatters, giving up their energy to the atoms and other electrons at each collision. The secondary photons, of lower energy than the originals, may continue to interact via the processes described above, and the positrons can encounter an electron and annihilate, producing gamma rays.

This section concentrates on attenuation, since we are most interested here in the transmission of the original beam through the biological tissue on to X-ray film or other detector; the scattered particles contribute to the fuzziness of the picture with a consequent loss of sharpness of the

X-ray image. We will be more interested in absorption when we come to study the biological effects of radiation in Chapter 5. However, we need to define some quantities that measure the absorption of X-rays before proceeding to a mathematical investigation of attenuation in §2.4.

2.3.5 *Units to measure the effect of radiation on matter*

As X-rays or gamma rays traverse matter, they cause ionization by one or more of the processes discussed above. In turn, the secondary particles can cause damage to the cells through which they pass. We need to define units to measure these effects.

It is convenient to measure the intensity of a beam of X-rays by allowing it to pass through a volume of dry air and collecting and measuring the charge produced. The resultant unit, called the **exposure,** is the **roentgen** (R), named after the discoverer of X-rays. An exposure of 1R will produce 2.58×10^{-4} C of positive charge in 1 kg of dry air at STP (density = 0.001293 g. cm^{-3}). Thus 1R will produce $(2.58 \times 10^{-4})/(1.6 \times 10^{-19}) = 1.6 \times 10^{15}$ ion pairs, since each ion carries the electron charge (an ion pair here means a charged positive ion and the ejected electron). This is one of the oldest units, still in common usage. Note that although the exposure of a beam of X-rays is directly proportional to the *number* of photons in the beam, it is not necessarily proportional to the *energy* of the photons; doubling the energy of each photon will not necessarily double the number of ion pairs, since other factors also influence the ability of photons to ionize atoms.

A more useful measure for studying the effects of radiation on matter is the amount of energy deposited by the radiation per unit mass of the material being irradiated. This is called the **absorbed dose** and the SI unit is **Grays** (Gy), named in honour of Louis Harold Gray (1905–1965). 1 Gray is defined as the radiation dose that deposits 1 Joule of energy in every kilogram of matter: 1Gy = 1 J/kg. An older unit of absorbed dose, still in use, is the **rad** (**r**adiation **a**bsorbed **d**ose), equal to 0.01 Gy (10 mGy).

Since the production of one ion pair requires an amount of energy that depends on the material (~1 eV to ~ 40 eV; 33.7 eV for air, close to 34 eV for all biological material), the dose is proportional to the exposure, with a constant of proportionality that depends on the material being irradiated. For dry air, we multiply the number of ion pairs produced by one roentgen in 1 kg of dry air by the energy required to produce each ion pair to obtain the energy deposited in 1 kg , or the absorbed dose, D_{air}. The result is $D_{air} = 8.69 \times 10^{-3} X$ Gy where X is the exposure in R, and D_{air} is in Grays (see Worked Examples, no.2). A useful rule-of-thumb follows: an exposure of 1R yields approximately 1 rad (10 mGy).

This discussion will continue in Chapter 4.

2.4 Attenuation of radiation

If a beam of radiation contains N particles in a cross-sectional an area S at right angles to the direction of the beam, the photon **fluence,** Φ, is defined by $\Phi = N/S$ — i.e. the number of particles that cross a unit area at right angles to the beam. (This quantity is sometimes called the flux density). Suppose a beam of X- or gamma rays, is incident on a sheet of material whose surface is at $x = 0$; let the fluence, the number of photons arriving at every square metre of the surface be $\Phi(0)$. These photons will interact with the atoms of the material (by one or other of the processes discussed in §2.3 above) and will be lost to the beam. This is the process we call attenuation of the beam.

Consider interactions at a depth in the material, x, where the fluence can be written as $\Phi(x)$; let dx be a small slice at this depth, x, (Figure 2.5) and let $dN(x)$, be the number of the photons at x that are lost to the beam by interaction or scattering by the atoms of the material contained in the slice between x and $(x+dx)$. The fraction of the number of photons lost in this slice is proportional to its thickness:

$$dN(x)/N(x) \infty\, dx = -\mu\, dx \qquad (2.1)$$

The constant of proportionality, μ, is called the **linear attenuation coefficient**, and the minus sign accounts for the fact that the original photons are being *lost* from the beam so that $N(x)$ is *decreasing*. Since $N/S = \Phi$ and $dN/S = d\Phi$, we can write $d\Phi(x)/\Phi(x) = -\mu\,dx$, or

$$d\Phi(x)/dx = -\mu\,\Phi(x) \qquad (2.2)$$

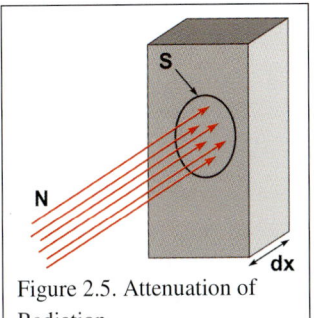

Figure 2.5. Attenuation of Radiation.

This differential equation has the solution:

$$\Phi(x) = \Phi(0)exp(-\mu x) \qquad (2.3)$$

as can be readily checked by differentiation to recover the original equation. $\Phi(0)$ is, as defined above, the fluence at $x = 0$, at the surface of the material.

The linear attenuation coefficient, μ, depends on the type and energy of radiation and the properties of the material. In particular, μ is proportional to the density of the material (ρ), as might be expected (but see Appendix A2.1), and it is often convenient to define the **mass attenuation coefficient** for a given material denoted by m, given the symbol $(\mu/\rho)_m$.

For dose calculations, both the 'density' of the beam, measured by the fluence, and the rate at which the particles arrive are important. We define the **fluence rate,** $\varphi(x)$, as the number of particles crossing a unit area in a unit time at a depth x in the material. Let $d\Phi(t)$ be the number of particles crossing the unit area at time t, in a time interval of dt. Then, by the definition of $\varphi(x)$ and using equation (2.3), we obtain:

$$\varphi(x) = d\Phi(x)/dt = [d\Phi(0)/dt] \times exp(-\mu x) = \varphi(0)exp(-\mu x) \quad (2.4)$$

As we might expect, both the fluence and the fluence rate decrease with distance as the beam traverses material.

The penetration of X-rays is much greater than that of electrons, as Figure 2.6 shows. This is a 'depth-dose' curve, which compares the penetrating power in tissue of 4 MeV X-rays to that of 12 MeV electrons; the scales have been normalized to make the dose agree at the maxima for each particle. (From their energy, you should be able to judge whether these X-rays would be used for therapeutic rather than diagnostic purposes). Note that even though the

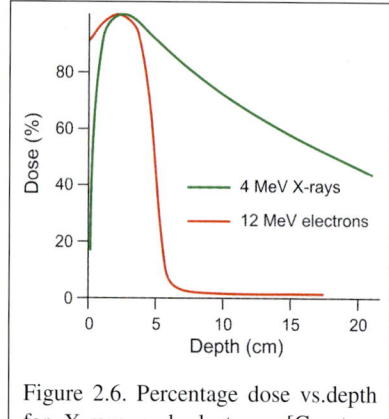

Figure 2.6. Percentage dose vs. depth for X-rays and electrons [Courtesy Oldham, 2001].

electrons are of much higher energy, they do not penetrate nearly so far into the tissue as do the X-rays.

2.5 Diagnostic use of X-rays

The use of X-rays for diagnostic purposes depends on the fact that the attenuation coefficient is different for bone, muscle, soft tissue and tumours. When X-rays pass through the human body, the reduction in the fluence depends on the variety of attenuation coefficients corresponding to the different types of material—tissue, muscle, or bone—encountered. If one type of material of thickness x_i has a linear attenutation coefficient μ_i, equation (2.3) is replaced by $\Phi(x) = \Phi(0) exp(-\sum \mu_i x_i)$ (2.5) Figure 2.7 shows the linear attenuation coefficients as a function of energy for some biological materials of interest. (Representative values of the corresponding mass attenuation coefficients are given in Appendix A2.1).

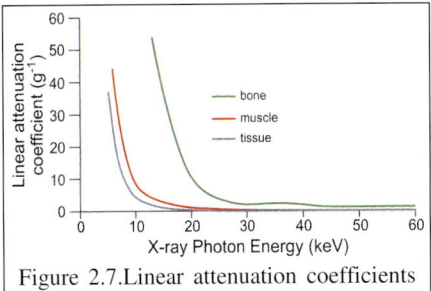

Figure 2.7.Linear attenuation coefficients for three biological materials as a function of photon energy [plotted from data in Hubbell and Seltzer, 2004, NIST; see Table A2.2].

Two features are immediately obvious; linear attenuation coefficients decrease rapidly with *decreasing* density of the material, from bone to muscle to tissue, and with *increasing* energies. The first effect provides good discrimination between bone and muscle or tissue; the discrimination is less good between muscle and tissue. The second feature ensures that the lower energy X-rays are preferentially attenuated as they pass into matter with a consequent 'hardening' of the beam. This feature is used to screen out the very low energy ('soft') X-rays that are not sufficiently penetrating to reach the X-ray film, but do deposit their unwanted energy in the patient. A thin foil of Aluminum is placed just after the target in diagnostic X-ray tubes in order to filter out some of this unwanted portion of the emitted spectrum.

The linear attenuation coefficient for air is about a thousand times smaller than that for biological tissue, thanks to the much lower density of air. Therefore, useful attenuation of X-rays can be achieved in air only by increasing the distance from the source (giving an $1/r^2$ reduction) or, most commonly, by inserting a thickness of heavy metal, with a large value of the linear attenuation coefficient as shielding.

The choice of the optimal X-ray energy for use in diagnosis must find a balance between using a low enough energy to provide good discrimination between bone, muscle, and tissue, and a high enough energy to allow the X-rays to penetrate these biological materials and reach the photographic plate or other detector. A typical spectrum of diagnostic X-rays covers a band of energies from ten to a hundred keV or so, while therapeutic X-rays have energies in the several MeV range, up to a maximum of around 25 MeV.

2.5.1 *X-ray images*

X-rays are a form of electromagnetic radiation like light; they are of higher energy, however, and can penetrate the body to form an image on film. Structures will appear as shades of gray on the photographic negative, depending on their density (bone appears white, corresponding to maximum attenuation, muscle and tissue appear gray, and air appears black). X-rays can provide information about obstructions, tumours, and other diseases.

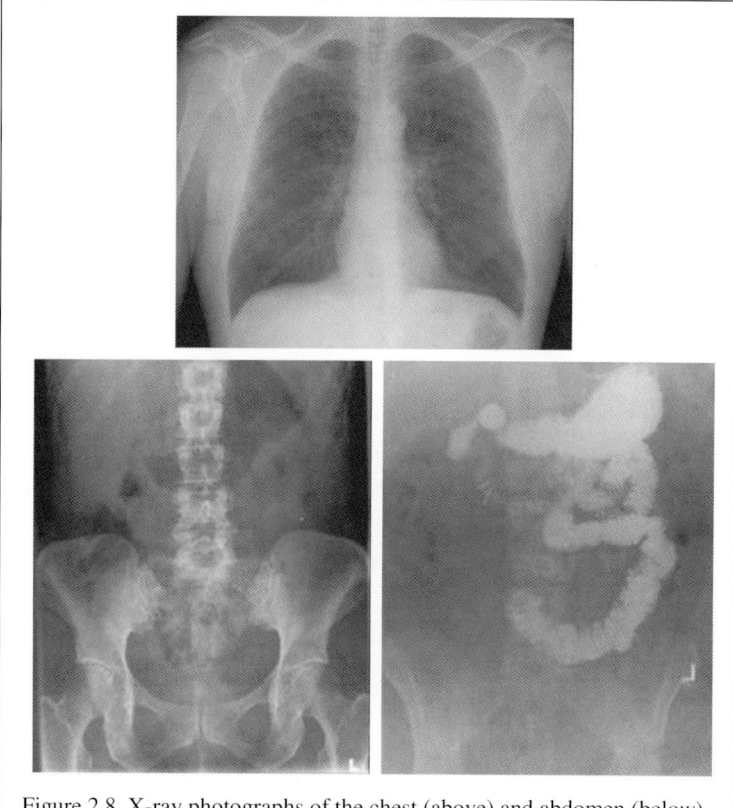

Figure 2.8. X-ray photographs of the chest (above) and abdomen (below).

The top picture in Figure 2.8 is a PA chest X-ray (so-called because the X-rays direction is from the Posterior (back) to the Anterior (front) of the body). The bones show up well, and some of the major arteries and blood vessels can be distinguished. Chest X-rays are usually PA, often supplemented by lateral views.

The lower two X-ray pictures in Figure 2.8 show the abdomen. The lighter parts of the image correspond to a greater attenuation of the X-ray beam. In the left one, the bones show up well but the soft tissues—giving a contrast of only several percent—are not well distinguished. Thus dyes or other more attenuating substances are often inserted or ingested to show up the soft tissue. The picture on the right was taken after the patient had been given a 'barium meal'; a previously ingested laxative has cleared out any other solid material. The stomach and lower bowel show up well. Nowadays, almost all X-ray images are digitized and viewed directly on a computer screen.

2.5.2 *Computer axial tomography (CAT, or CT scans)*

'Tomography' means the graphical representation of slices; the 'axial' refers to the plane at right angles to the central axis of the body (to be correct, it should be called 'trans-axial'). In a standard X-ray exposure, the images of all the structures in the path of the X-rays fall on top of each other in the photographic image, as is clear from Figure 2.8.

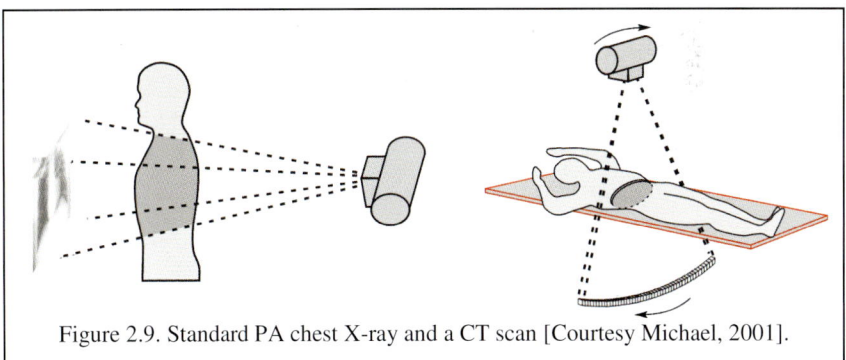

Figure 2.9. Standard PA chest X-ray and a CT scan [Courtesy Michael, 2001].

To obtain information about the different structures, Godfrey Hounsfield in 1972 pointed out the advantage of taking many exposures at a variety of different angles. (Hounsfield and Allen Cormack shared the 1979 Nobel Prize in Physiology or Medicine.) These can then be combined using sophisticated computer programmes to create a 'picture' of 2-dimensional slices of the exposed volume. To make reconstruction computationally possible the photographic film is replaced by many small solid state detectors. The CT scanner on the right of Figure 2.9 has a beam of X-rays that spread out and a small number of detectors. Figure 2.10 indicates how many projections can be used to provide a 2-dimensional image of a slice of the body.

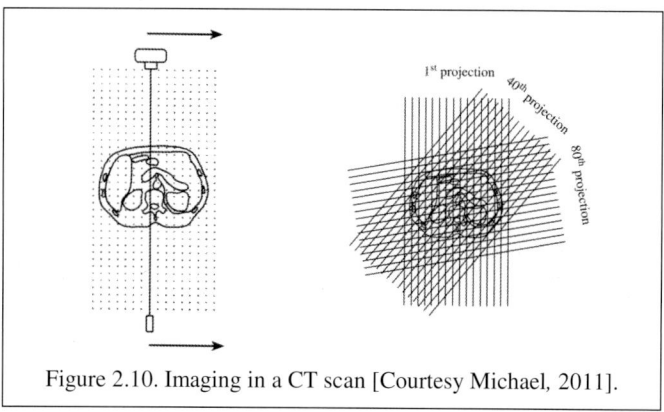

Figure 2.10. Imaging in a CT scan [Courtesy Michael, 2011].

Modern CT scanners are now much more sophisticated. Figure 2.11 shows a '4th generation' scanner that involves a moving X-ray tube and a circular array of detectors that surrounds the patient. In helical computed tomography, the patient is moved through the X-ray beam as shown on the right, so that many slices can be taken in a short time.

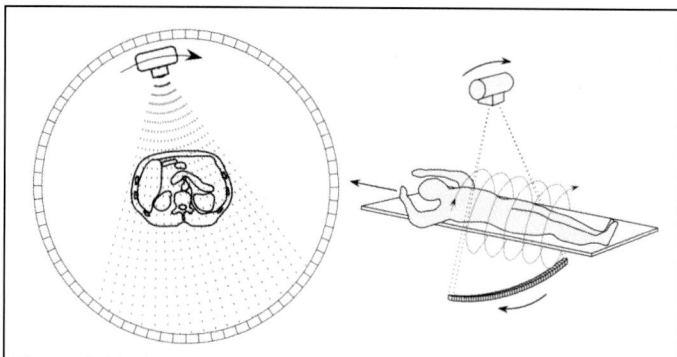

Figure 2.11. On the left: fourth generation scanner. On the right: helical tomography [Courtesy Michael, 2001].

Figure 2.12 shows the CT scan of the abdomen, corresponding to the abdominal slice of the patient shown in figures 2.9 to 2.11. The liver, the spinal column, and the kidneys can all be clearly seen.

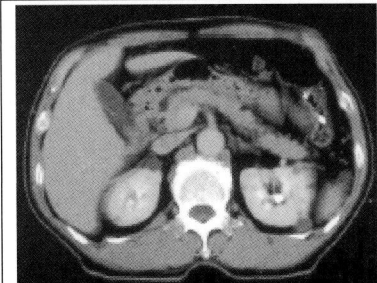

Figure 2.12. CT scan of the abdomen [Courtesy Michael, 2001].

When the different 2-dimensional images are combined, a sophisticated computer calculation can provide what amounts to a 3-D reconstruction. (Similar computational techniques are used in Magnetic Resonance Imaging, Chapter 6).

A2.1 *Some useful physical quantities*

Note that bone is about twice the density of tissue or muscle, and that the electron density (given by ZN_A/A) is approximately constant across the periodic table.

Table A2.1. Useful physical quantities

Material	Density (g cm^{-3})	Effective Atomic Number	No. of Electrons per gm. (ZN_A/A) ($\times 10^{-23}$ g^{-1})
Air	0.001293	7.6	3.01
Water	1.00	7.4	3.34
Muscle	1.00	7.4	3.36
Tissue	0.95	5.9	3.48
Bone	1.85	13.8	3.00
Lead	11.7	82	2.38

Table A2.2 gives the mass attenuation coefficients for various materials and energies. The linear attenuation coefficients obtained by multiplying these mass attenuation co-efficients by the approp-riate density, are plotted in Figure 2.7

Table A2.2. Mass attenuation coefficients $(\mu/\rho)_m$ (cm^{2}g^{-1}) [NIST, Hubbell and Seltzer, 2004]			
Energy (keV)	Tissue	Muscle	Bone
10	3.268	5.356	28.51
20	0.568	0.821	4.000
40	0.239	0.269	0.666
60	0.197	0.205	0.315
80	0.180	0.182	0.223
100	0.169	0.169	0.186
200	0.137	0.136	0.131
500	0.097	0.096	0.090
1000	0.071	0.070	0.066
2000	0.049	0.049	0.046
5000	0.030	0.030	0.029
10000	0.021	0.022	0.023
20000	0.017	0.018	0.021
Density (g/cm^3)	0.95	1.00	1.85

A2.2 *Photon scattering and attenuation coefficients*

The probability of scattering depends on the **cross-section for scattering** of the atoms that compose the irradiated material. It is easiest to think of this cross-section as the area that the atom presents to the incident beam for the particular type of scattering under consideration: denote it by σ (pronounced 'sigma'). Consider a very thin slice of the material, of thickness dx, on to which a beam of photons of cross-sectional area S impinges. The probability of scattering, P, is then equal to the total area of the cross-sections of the n atoms presented to the beam, divided by the total area of the beam. (The choice of a very thin slice of material ensures that no atom or electron is hidden behind others.)

Thus $P = n\sigma/S$

(A2.1)

Now n, the number of scattering centres encountered by the beam in the area S, is given by the number of atoms per mole, N_A, multiplied by the number of moles in the volume of the slice being irradiated, equal to $(\rho S dx)/A$, where ρ is the density of the material and A is its mass number.

$$P = (N_A)(\rho S dx / A)(\sigma / S) = (N_A/A)\sigma \rho dx \qquad (A2.2)$$

P is equal to the fraction of photons in the beam that are scattered, dN/N.

$$P = |dN/N| = \mu \, dx \qquad (A2.3)$$

The second equality is Equation (2.1) in §2.4.
Equating equations (A2.2) and (A2.3), we obtain:

$$\mu = (N_A/A)\sigma \rho \qquad (A2.4)$$

and the mass attenuation coefficient:

$$\mu/\rho = (N_A/A)\,\sigma \qquad (A2.5)$$

Now we are in a position to consider the factors that determine how an X-ray beam is attenuated by different physical processes.

The Photoelectric Effect   At X-ray energies, where the photoelectric effect dominates, quantum mechanical calculations show that the probability of interaction per atom is approximately proportional to $Z^4 E_\gamma^{-3}$, where Z is the atomic number of the atom, and E_γ is the energy of the X-ray. Thus we have, approximately, $\sigma \propto Z^4 E_\gamma^{-3}$, and:

$$\mu = (N_A/A)\sigma \rho \propto (N_A/A)Z^4 E_\gamma^{-3}\rho = (Z N_A/A)(Z^3 E_\gamma^{-3})\rho \qquad (A2.6)$$

Since (ZN_A/A) is approximately constant across the periodic table,

$$\mu \propto Z^3 \rho \qquad (A2.7)$$

for a given energy. The beam attenuation for a given energy is given by:

$$\Phi(x) = \Phi(0)\exp(-\mu x) \propto \Phi(0)\exp(-Z^{-3}\rho x) \qquad (A2.8)$$

The Compton Effect. At higher energies, where Compton scattering becomes important, the scattering probability per atom depends approximately on the number of electrons available for scattering. Thus σ is directly proportional to Z :

$$\mu = (N_A/A)\sigma\rho \propto (N_A/A)Z\rho \ = \ (ZN_A/A)\rho \tag{A2.9}$$

Thus at a given energy, the beam attenuation is given by:

$$\Phi(x) = \Phi(0)\,exp(-\mu x) \propto \Phi(0)\,exp(-\rho x) \tag{A2.10}$$

As expected, for a given thickness of material at a given energy, the attenuation increases rapidly with density.

Since ZN_A/A is approximately constant across the periodic table, equation (A2.9) shows that the mass attenuation coefficient, μ/ρ, is approximately equal for bone, muscle and tissue at energies where Compton scattering dominates (you can check this prediction in the Table A2.2 above). The linear attenuation coefficient, μ, is directly proportional to density. Thus, except at the lowest energies of this range, the ability of the X-rays to discriminate is dominated by the density differences of the biological materials. At the lowest X-ray energies, where the photoelectric effect dominates, this density-dependence is greatly enhanced, since the density of a material is closely dependent on the atomic number, Z , and $\mu \propto Z^3\rho$ (Equation A2.7).

Thus the best contrast between bone and tissue occurs at lower X-ray energies.

Exercises Chapter 2

1. Old fashioned TV images are produced by electron striking a fluorescent screen. What is the short wavelength limit of the radiation emitted when the electron is brought to rest on the screen after being accelerated through a tube voltage of 5 kV? In what region of the electromagnetic spectrum does this radiation lie?
2. X-rays of wavelengths greater than 0.1 nm are required for a specific medical procedure. What is the minimum value of the tube voltage that will produce X-rays of the required wavelength?

3. Radiation delivers energy to the tissue it penetrates. It might be thought that the damage it causes could be caused by heating. To test this hypothesis, consider the following. An X-ray dose of 2 Gy can be lethal. If the equivalent energy were absorbed as heat, what would be the rise in body temperature? (The specific heat of tissue may be assumed to be that of water: 4200 $J.kg^{-1}.K^{-1}$)

4. An X-ray machine has a target made of lead. Two main characteristic lines show up in the X-ray spectrum. A schematic energy level diagram showing the two relevant transitions is shown. Calculate the wavelength, in pm, of the least energetic photon in this characteristic line spectrum. What is the minimum tube voltage that is required to produce this spectrum?

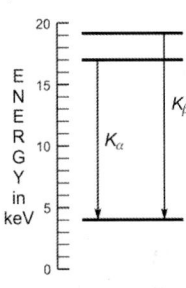

5. A photon of energy 200 keV scatters off a free, stationary electron. If the energy of the scattered electron is 60 keV, what is the wavelength of the scattered photon. What is this process called?

6. An X-ray photon of wavelength 4.9 pm enters tissue and undergoes a Compton scattering. The scattered photon has a wavelength of 6.4 pm. What is the energy of the electron produced in this process?

7. A photon of frequency 2.2×10^{19} Hz is totally absorbed by an atom, which subsequently emits an electron. What is the electron's energy? (You may assume that the binding energy of the electron to the atom and any resultant excitation of the atom are negligible). What is this process called?

8. Calculate the number of ion pairs produced by 1.5R of X-rays in 0.5 litres of air.

9. The thickness of tissue required to reduce the flux density of a beam of X-rays to half its original value is called the Half Value Layer (HVL). Using the relationship for photon fluence as a function of depth, derive an expression for the HVL in terms of the linear attenuation coefficient, μ.

10. A chest X-ray is shown in Figure 2.8. If the X-rays used to take the X-ray can be assumed to have an average energy of 60 keV, what, approximately, would you expect to be the ratio of intensities on the

photograph for X-rays that traverse 1.5 cm of a rib compared to the X-rays that traverse 10 cm of lung tissue? (Assume that air filled lung tissue has a linear attenuation coefficient of 0.060 cm^{-1}). Does your result seem reasonable?

11. In order that a diagnostician can see reasonable images on an X-ray film, an exposure of not more than approximately 20 mR must be delivered to the film during exposure (remember that X-ray images—see Figure 2.8—are negatives, producing more exposure of the film, and thus a darker image, when the X-rays are reduced by their passage through body tissue than when the X-rays pass through muscle and bone). A) For 60 keV X-rays, what is the exposure (in R) needed to reach the required density of the exposed film? (Take the width of an 'average' person to be 25cm. You will also need to make a fairly obvious assumption about the reduction of the exposure as the X-rays traverse the body tissue.) B) What would be the required exposure for 40 keV X-rays? C) Without doing the calculation, at which energy would you expect the absorbed dose to the patient to be greater?

12. The radioisotope ^{99m}Tc emits gamma rays of 0.140 MeV. It is known that the interactions of ^{99m}Tc gamma rays in the patient occur primarily by the Compton process, whereas those in the sodium iodide crystal used in a 'gamma' camera (*guess what this camera detects?*) are mainly photoelectric processes? Give a plausibility argument as to why this should be so.

Chapter 3

Radioactivity and Radioisotopes

3.1 Introduction

We have seen in Chapter 1 that many nuclei have isotopes (with the same proton number, Z, but different neutron numbers, N, and mass numbers $(A = N+Z)$. The emissions first discovered by Becquerel in 1896 (§1.2) were the result of decays of unstable isotopes (radioisotopes) of Uranium present in rock samples. Becquerel jointly won the 1903 Nobel Prize in Physics with Marie and Pierre Curie, who had isolated and identified radium from pitchblende, a kind of tar. In 1911 Marie also won the Nobel Prize in Chemistry. Several years were to pass before the nature of these radioactive emissions were untangled.

Naturally occurring radioactive nuclei must either be almost stable (if they did not have a very long decay time they would no longer exist!), or they must be continuously created by natural processes. For instance, ^{14}C, a radioisotope of the stable ^{12}C, is continuously produced by the nuclear interactions of cosmic rays with the atoms of the upper atmosphere. Most radioactive nuclei used for diagnostic or therapeutic purposes are created artificially at nuclear reactors or particle accelerators.

3.2 Radioactive decay

The decay of nuclei is a random process governed by the probabilistic nature of the quantum world. For each nucleus of a specific radioisotope the probability of decay at any instant of time is the same, though we cannot determine ahead of time which nucleus it will be. However, with a large enough number of nuclei, the number that decay in a given time can be closely approximated.

Suppose that a sample of a radioisotope has $N(t)$ nuclei at time t. The number of nuclei that decay in any given interval of time between t and $t+dt$ is proportional to the length of the time interval dt and to the number of nuclei present, $N(t)$.

$$dN(t) = N(t + dt) - N(t) = -\lambda N(t)dt \qquad (3.1)$$

where λ, the constant of proportionality, that depends on the particular nucleus concerned, is called the **decay constant**. We have already seen a solution to this linear differential equation in §2.4 (Equations 2.2 and 2.3): the solution here is

$$N(t) = N(0)exp(-\lambda t) \qquad (3.2)$$

where $N(0)$ is the number of nuclei present at time $t = 0$.

It is useful to define the **half-life**, $T_{1/2}$, the time in which half of the original nuclei remain. Setting $N(t) = N(0)/2$ $\qquad (3.3)$
for $t = T_{1/2}$ in equation (3.2), the half-life

$$T_{1/2} = \ln_e 2/\lambda = 0.693/\lambda \qquad (3.4)$$

Thus in one half-life, only 50% of the nuclei remain; after another half-life only 25% of the original number remain, and so on.

A more directly measurable, and therefore more useful measure is the **activity** (or the Decay Rate) of a sample, which is the number of decays per second at time t, often denoted by the symbol $R(t)$:

$$R(t) = |dN/dt| = \lambda N(0)exp(-\lambda t) = R(0)exp(-\lambda t) \qquad (3.5)$$

$R(0)$ is obviously the activity at time $t = 0$.

Thus we see that:
i)   R(t) follows the same exponential decay law as $N(t)$ and,
ii)   that $R(t)$ and $N(t)$ are related by $R(t) = \lambda N(t)$ $\qquad (3.6)$

The older unit of activity is the **Curie (Ci),** which is defined to be the number of decays per second of 1g of Radium, or **3.7 x 10^{10}** decays per second. The SI unit is the **Becquerel (Bq)** which is defined as 1 decay per second. Thus **1 Ci = 3.7 x 10^{10} Bq**.

3.2.1 *Types of decay*

There are four mechanisms of radioactive nuclear decay. In each case, a **parent** nucleus decays into a **daughter** nucleus and another particle or particles. The total number of nucleons (protons and neutrons) is always conserved.

3.2.2 *Alpha decay*

This was the first mode of decay identified by Ernest Rutherford, who showed that the radioactive emission, called the alpha particle, was the nucleus of the helium atom, $^4He^{++}$ (§1.2). Thus the decay can be written:
$$_Z^A X \rightarrow \, _{Z-2}^{A-4} Y + \, _2^4 He.$$

There is an apparent impossibility in alpha decay that had to await the advent of quantum mechanics for its resolution. Typically, alpha particles have kinetic energies of around 5 MeV. For the alpha particle to emerge from inside the nucleus, it must pass through a region where its Coulomb potential energy is typically 20 or 30 MeV.

To appreciate why this is (classically) impossible, consider the reverse process—the approach of an alpha to the daughter nucleus. Alpha particles from radioactive decay may be treated as classical Newtonian (non-relativistic) particles. A long way from the nucleus the total energy of the alpha particle is equal to its kinetic energy, and its potential energy in the Coulomb field of the nucleus is zero. As it approaches the (positively charged) nucleus, the repulsive Coulomb force will gradually slow it down. In other words, some of its

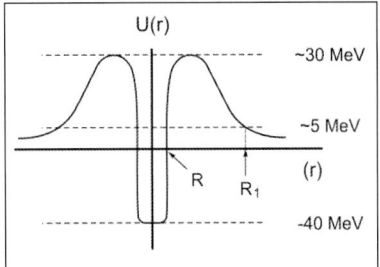

Figure 3.1. Potential well seen by an alpha particle inside a nucleus.

kinetic energy will be transformed into Coulomb potential energy; think of a ball being rolled up a hill under the action of gravity. When the alpha comes to rest, at R_I (> the radius of the nucleus, R) in Figure 3.1, all of its energy is potential. It cannot approach more closely to the nucleus where the potential energy is higher, because otherwise its kinetic energy would have to become negative, which is (classically) impossible. In the 'reverse' process of alpha decay, the alpha, which presumably has a kinetic energy of 5 MeV inside the nucleus where the nuclear force holds sway, would, according to classical mechanics, be unable to move away from the core of the nucleus (where the Coulomb force is zero and the nuclear force dominates) into a region where its kinetic energy would be negative.

We now know that matter (like light, see A1.2.) displays both particle-like and wave-like behaviour. The wave-like nature of the alpha allows it to 'leak' through the Coulomb potential barrier with a small, but finite, probability. As might be expected on 'common-sense' arguments, this probability increases with increasing alpha energy and decreasing size of the nucleus. The measured lifetimes of alpha emitters confirm this expectation.

3.2.3 *Beta decay*

In 1899, Rutherford separated the newly discovered emissions from nuclei into alpha decay (§1.2) and beta decay. In beta decay, the nucleus emits electrons or positrons (the positively charged anti-electron). Beta decay thus is either

$$^A_Z X \rightarrow \,_{Z+1}^{A}Y + e^{-1} \text{ or } ^A_Z X \rightarrow \,_{Z-1}^{A}Y + e^{+1} \tag{3.7}$$

A typical spectrum of the kinetic energy of the beta rays from such decays is shown in the figure: the spectrum covers a wide range from low values up to a maximum, called the endpoint energy, K_{max}. This poses a theoretical problem, since, in a decay in which there are only two decay products, momentum and energy conservation yield momenta and kinetic energies for each that are fixed by the energy available for the decay (two equations—energy and

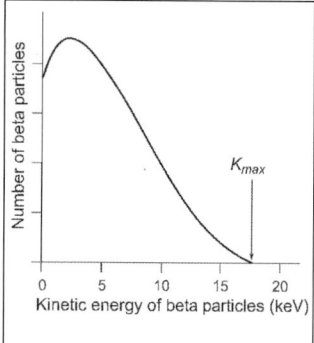

Figure 3.2. Energy spectrum of a beta emitter.

momentum conservation—and two unknowns—the momenta of the two decay products). The electron's energy, calculated on the assumption of the above two-body decays, yields a value that is equal to the observed endpoint energy, K_{max}; the observation of the lower energy continuous spectrum means that more than two decay products are present, or that energy is not conserved.

Accordingly, in 1930 Wolfgang Pauli argued that beta decay must involve three particles, not two, since the alternative explanation—non-conservation of mass-energy—was unthinkable. He proposed that the three were the daughter nucleus, the beta particle (electron or positron) and a new, hitherto unobserved neutral particle which Fermi dubbed the neutrino (little neutral one). Then the maximum energy of the beta particle corresponds to the case in which the neutrino is emitted with zero kinetic energy. From the equality of the endpoint energy and the energy calculated on the assumption that the electron and daughter are the only decay products, it is clear that the neutrino mass must be very small; in our calculations we can assume that the neutrino mass is zero. (We now know that there are three different types of neutrino, with very small, but non-zero masses!). The other observed energies in the spectrum correspond to decays in which the neutrino carries off some of the available kinetic energy.

Denoting the neutrino by ν (pronounced "new") we rewrite the beta decays of equation (3.6) by:

$$^A_Z X \rightarrow _{Z+1}^A Y + e^{-1} + \nu \text{ or } ^A_Z X \rightarrow _{Z-1}^A Y + e^{+1} + \nu \qquad (3.8)$$

(In common with all other nuclear particles, the neutrino has an antiparticle; the neutrino accompanies the e^+ decay and the antineutrino accompanies the e^- decay.)

3.2.4 *Gamma decay*

In 1900, the Frenchman Paul Villard detected a radiation emitted from radium. Three years later, Rutherford realized that this type of radiation was different from the two he had previously identified, and named it gamma radiation. Just as excited atoms can decay with the emission of electromagnetic radiation, so can excited nuclei: the decay is written $^A_Z X^* \rightarrow$ $^A_Z X + \gamma$. The excited nuclei can be produced as the daughter of a previous alpha or beta decay, or in nuclear interactions with other nuclei.

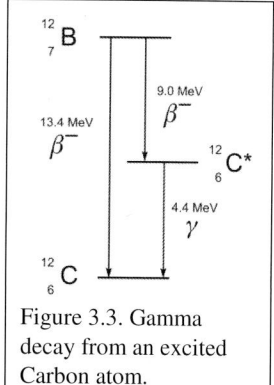

Figure 3.3. Gamma decay from an excited Carbon atom.

Often two or more gammas are emitted as successive energy levels are reached. Figure 3.3 shows a gamma decay from an excited level of ^{12}C—the excitement is denoted by * !

Although gamma decay is usually very rapid (around 10^{-15} seconds), in rare cases the excited state lasts for much longer (usually for hours, very occasionally for years). The excited state is then called an **isomeric state** and the excited state and the final stable ground state are called **nuclear isomers**. These long-lived isomeric states are often labeled with a small 'm' to indicate their metastable nature (e.g. technetium, ^{99m}Tc, which, as we shall see in §3.5.1 is used in diagnostic medicine).

3.2.5 *Electron capture*

In electron capture (EC), often classified with beta decay for obvious reasons, the nucleus captures an orbital electron (most usually from the K shell), one of the protons in the parent is converted to a neutron, and a neutrino is emitted $^A_Z X + e^- \rightarrow {}_{Z-1}^{A} Y + \nu$. X-rays are emitted as electrons in the atom cascade down to fill the vacancy created in the lower shell. These X-rays find application in radiotherapy (§3.5, Table 3.1).

3.3 The energetics of radioactive decay

The mass-energy equivalence is beautifully manifested in radioactive decay. The mass of the parent is always greater than the sum of the masses of the daughter and the emitted particle or particles; the difference in mass is transformed into the kinetic energy of the decay products (via $E = mc^2$). This is called the **disintegration energy**, or Q **value** of the decay.

In alpha decay, the Q value is equal to the total kinetic energy ($K.E.$) carried off by the daughter and the alpha, both of which have the same momentum. Since the momentum p of a mass m is related to the kinetic energy by $K.E. = p^2/2m$, (at these energies, classical non-relativistic mechanics holds), the kinetic energy of the alpha is far greater than that of the daughter. Thus, to a good approximation, the kinetic energy of the alpha can be equated to the Q value.

In beta decay, the Q value again equals the total kinetic energies of the decay products. If we set the neutrino kinetic energy to zero, and again neglect the kinetic energy of the much more massive daughter, the endpoint beta energy is then just equal to the Q value: $K_{max} = Q$.

In gamma decay, the gamma ray energy ($h\nu$) is equal to the difference in energies between the initial and final energy levels of the nucleus, if we make the excellent approximation that the recoil kinetic energy of the daughter is zero. Alternatively, the difference in mass between the excited parent nucleus and the mass of the daughter, expressed as energy, almost exactly equals the gamma ray energy.

The Q value in electron capture equals the kinetic energies of the daughter and the neutrino; as usual, that of the former is negligible. Since the neutrino interacts hardly at all with matter, we are usually interested in the effects of the emitted X-rays, the energies of which depend on the atomic makeup of the parent nucleus.

There is one odd wrinkle to all of these calculations. The masses required for the calculations are, of course, the nuclear masses; however it is the atomic masses of the corresponding atoms that are susceptible to precise measurements, and are the masses provided in atomic mass tables. The required nuclear masses must be calculated from the provided atomic masses by subtracting Z electron masses; the binding energies of these Z electrons can be assumed to be negligible.

3.4   The generation of radioisotopes

Most radioisotopes are created artificially by bombarding stable nuclides with beams from particle accelerators or by neutrons from nuclear reactors. Fission products from reactors are also a rich source of radioisotopes. Many modern medical facilities have on-site cyclotrons to produce their own radioisotopes. Chalk River Laboratories in Ontario were once leading players in this industry.

Many radioisotopes are produced by **generators.** This section discusses the method of production for the commonly used radioisotope, technetium, ^{99m}Tc.

Consider a radioactive decay chain, in which a parent radioisotope decays to a daughter. Let $N_1(t)$ be the number of nuclides of the parent at time t, with half-life T_1 and $N_2(t)$ be the number of daughter nuclides at time t, with half-life T_2. Then the rate of change of the daughter can be written as:

$$
\begin{aligned}
(d\,N_2(t)/dt) \\
&= (rate\ of\ formation\ of\ N_2) - (rate\ of\ decay\ of\ N_2) \\
&= (rate\ of\ decay\ of\ N_1) - (rate\ of\ decay\ of\ N_2) \\
&= \lambda_1 N_1(t) - \lambda_2 N_2(t) \qquad\qquad (3.9)
\end{aligned}
$$

Substitute $N_1(t) = N_1(0)exp(-\lambda_1 t)$ from equation (3.2), and the equation can be solved to give an expression for the activity of the daughter, $R_2(t)$, in terms of that of the parent, $R_1(t)$ (see Appendix A3.1):

$$
\begin{aligned}
R_2(t) = \lambda_2 N_2(t) &= \lambda_1 N_1(t)\{\lambda_2/(\lambda_2 - \lambda_1)\}\{1 - exp[-(\lambda_2 - \lambda_1)t]\} \\
&= R_1(t)\{\lambda_2/(\lambda_2 - \lambda_1)\}\{1 - exp[-(\lambda_2 - \lambda_1)t]\} \qquad (3.10)
\end{aligned}
$$

An interesting situation arises when the parent has a half-life that is much greater than that of the daughter: $T_1 \gg T_2$ or $\lambda_1 \ll \lambda_2$. In that case the activity of the daughter is given by:

$$R_2(t) = R_1(t)\{1 - exp(\lambda_2 t)\} \qquad (3.11)$$

This function is plotted in Figure 3.4. At times long compared to T_2, the activity of the daughter becomes equal to the activity of the parent: $R_2(t) = \lambda_2 N_2 = \lambda_1 N_1 = R_1(t) \qquad (3.12)$

This is called '**secular equilibrium**'.

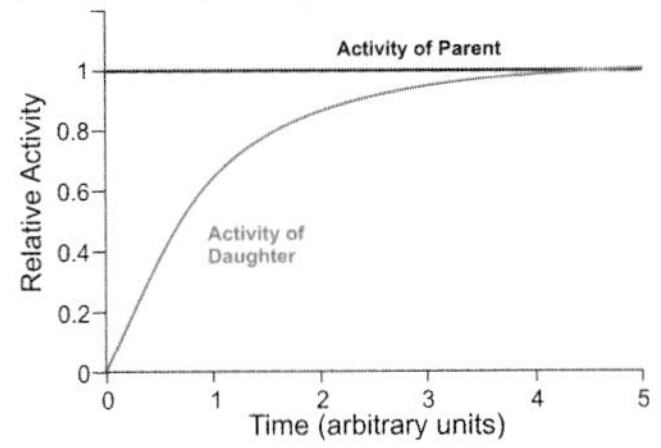

Figure 3.4. Generation of daughter radioisotope.

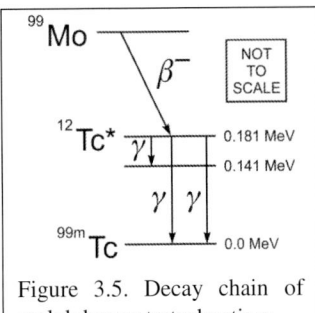

Figure 3.5. Decay chain of molybdenum to technetium.

For the production of ^{99m}Tc, the parent is ^{99}Mo (molybdenum–99). A simplified version of the relevant decay chain is sketched in Figure 3.5. The half-lives of ^{99}Mo and ^{99m}Tc are about 2.5 days and 6 hours respectively.

The ^{99}Mo, produced as a fission product or by neutron bombardment of ^{98}Mo, is absorbed on alumina. Assuming that all of the ^{99}Mo decays to ^{99m}Tc, the latter accumulates according to the equation (3.10)*. At regular intervals the column is flushed with a saline solution, which dissolves the ^{99m}Tc, leaving the Mo behind to generate more of it. In this way the ^{99}Mo acts as a 'cow' that can be 'milked' (the fancy word is 'eluted') of the more useful ^{99m}Tc.

---

* In fact, since only 0.88 of the ^{99}Mo decays to ^{99m}Tc, $R_2(t)$ has to be multiplied by this fraction for this particular decay.

The activities of the two radioisotopes when the Mo cow is milked every day are shown in Figure 3.6. Note that both parent and daughter decay, and that about a day (approximately 4 half-lives) is required for the ^{99m}Tc to reach its (approximately) maximum activity after milking. When that happens, '**transient equilibrium**' is said to have been reached.

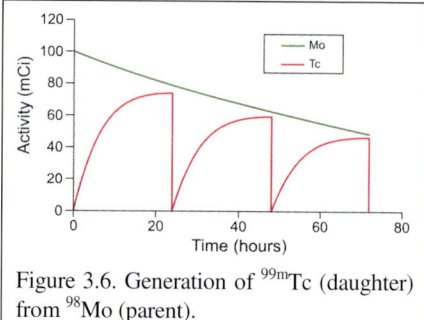

Figure 3.6. Generation of ^{99m}Tc (daughter) from ^{98}Mo (parent).

3.5 Diagnostic use of radioisotopes

Radioisotopes have a multitude of uses in both diagnosis and therapy. These notes will touch on only a few of the interesting topics. The therapeutic use of radioisotopes is discussed in Chapter 4.

For diagnostic use, radioisotopes are attached to a drug or a pharmaceutical that is known to target a specific organ. The pharmaceutical is then introduced into the patient by injection, inhalation, or ingestion, and its subsequent distribution is observed by detecting the emitted radiation. The resultant data is analyzed using equally sophisticated software. In this case, the radioisotope acts as a tracer for a specific physiological process. Diagnostic nuclear medicine examinations are used to identify abnormalities in the brain, thyroid, heart, lung, kidney, liver, spleen, and bone.

Table 3.1. Half Lives of some common radioisotopes[LBNL]

Radio isotope	Half-life	Decay Mode	Organ to be scanned
^{123}I	13 hours	EC , γ	Thyroid
^{131}I	8 days	β^- , γ	Thyroid
^{198}Au	2.7 days	β^- , γ	Liver
^{201}Tl	3.0 days	EC , γ	Heart
^{111m}In	2.8 days	IT, γ	Blood
^{85}Sr	65 days	EC , γ	Bone

EC: Electron Capture.
IT: Isomeric Transition

3.5.1 *Tracers*

A compound that concentrates in the organ of interest is tagged by attaching a known radioisotope. Technetium (^{99m}Tc) is one of the most commonly used in scans of many different organs. It combines with many chemical compounds, and its half-life of 6 hours is short enough to keep the long-term dose low, but long enough to allow adequate time for a good signal. Some other commonly used radioisotopes are listed in Table 3.1. Typical tracer activities are around 0.1 mCi.

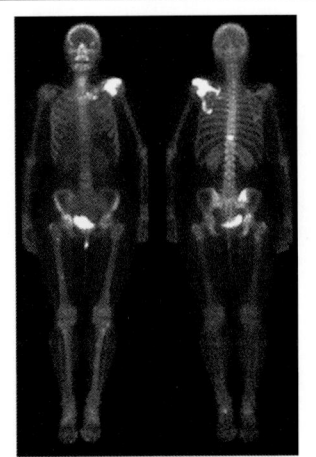

Figure 3.7 shows posterior and anterior whole-body scans of a patient who has ingested ^{99m}Tc. The 'hot spots' indicate disease in the bone. The difference in intensity of the hot spots is due to the attenuation of the radiation in the patient's bone and tissues.

Figure 3.7. Whole-body scan of a patient who has ingested the radioisotope ^{99m}Tc [Courtesy, Badawi, 2001].

3.5.2 *Single Photon Emission Computed Tomography*

In single photon emission computed tomography (SPECT), the gamma radiation from a radioisotope in the patient's body is detected at a variety of angles around the patient, and the results used to reconstruct slices through the patient using a process similar to that used for CT scans. The dose absorbed by the patient is greater than that absorbed during a single planar view, due to the longer time required for a SPECT scan.

3.5.3 *Positron Emission Tomography (PET)*

PET uses the pair production that occurs after a positron, emitted from an ingested radiopharmaceutical, annihilates with an electron in the body. The positron emerges from the decay and slows down in tissue within a few millimeters; when it is virtually at rest, it annihilates with a local electron. The two resultant 511 keV gamma rays (the electron and positron have the same mass of 511 keV/c^2), conserving momentum, travel in directly opposite directions out of the patient's body. Simultaneous (coincident) detection of the two gamma rays gives a precise line along which the radionuclide must lie. A measurement of the difference between the times of arrival of the two photons yields the position along this line. The detection of many such coincident events at different angles allows the area where the radiopharmaceutical has concentrated to be precisely located. The experimental set-up is similar to that used for CT scans discussed in Chapter 2.

Table 3.2. PET Radioisotopes [LBNL].

Radio isotope	Half-life (min)	Max. β+ Energy (MeV)
^{11}C	20	0.96
^{13}N	10	1.19
^{15}O	2	1.70
^{18}F	110	0.64
^{82}Rb	1	3.15

Some of the radionuclides available for PET are listed in Table 3.2. All are all very short lived, so an on-site cyclotron is a necessity for all but ^{18}F, the most widely used.

Figure 3.8 shows a schematic of the detector array and indicates how the coincident signals are used to remove background noise.

Fluoro-deoxyglucose (FDG), in which the ^{18}F has been incorporated, is absorbed into cells that have a high glucose metabolic rate—such as tumours. Once it is in the cell it does not further metabolize, so it accumulates in tumours where it serves as an excellent marker. This type of tomography is particularly valuable for **functional analyses**—i.e. investigations in which biological functions can be studied as they happen.

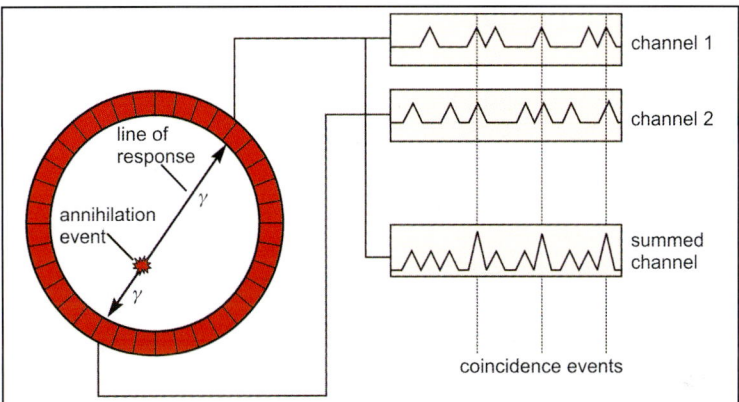

Figure 3.8. Detector array used in PET scans. The coincidence between the two opposing gamma rays allows a selection of the signal from the background noise [Courtesy Badawi, 2001].

3.5.4 *Isotopic dilution*

Radioactive tracers can be used to determine volume of a fluid in the body. A radiopharmaceutical with a known activity, R, is introduced to the body. Once the radioisotope has dispersed throughout the bodily fluid, of total volume, V, a small volume, v, of fluid is extracted. Its activity, r, is then measured. The total volume of the fluid that received the radiopharmaceutical is then given by $V = vR/r$ (3.13)

This calculation assumes that the isotope has been thoroughly mixed in the fluid, and that there has been no dilution by nuclear or biological excretion (see §4.5.3).

Similarly, dilution techniques can measure the amount of what is called the **exchangeable pool** of a given element in the body; this is that part of the element being measured that exchanges with the environment through ingestion and excretion. For example, the exchangeable calcium in the body can be measured by dilution techniques; obviously the calcium deposited in the bones is not part of this exchangeable pool.

A3.1 *Calculation of parent –daughter generation*

From Equations (3.2) and (3.9),

$$dN_2(t)/dt = \lambda_1 N_1(t) - \lambda_2 N_2(t) = \lambda_1 N_1(0)exp(-\lambda_1 t) - \lambda_2 N_2(t)$$

$$(A3.1)$$

Now search for a solution: $N_2(t) = K\{exp(-\lambda_1 t) - exp(-\lambda_2 t)\}$
where K is a constant. This satisfies the requirement that $N_2(t) = 0$, when
$t = 0$ and $t = \infty$.
Inserting this trial solution Equation 3.12 yields:

$$LHS = K\{-\lambda_1 exp(-\lambda_1 t) + \lambda_2 exp(-\lambda_2 t)\}$$
$$= -K\lambda_1 exp(-\lambda_1 t) + K\lambda_2 exp(-\lambda_2 t) \qquad (A3.2)$$

$$RHS = \lambda_1 N_1(0)exp(-\lambda_1 t) - \lambda_2 K\{exp(-\lambda_1 t) - exp(-\lambda_2 t)\}$$
$$= \lambda_1 N_1(0)exp(-\lambda_1 t) - \lambda_2 Kexp(-\lambda_1 t) + K\lambda_2 exp(-\lambda_2 t) \quad (A3.3)$$

Equating yields:

$$-K\lambda_1 exp(-\lambda_1 t) = \lambda_1 N_1(0)exp(-\lambda_1 t) - \lambda_2 Kexp(-\lambda_1 t)$$
Or: $K = \{\lambda_1/(\lambda_2 - \lambda_1)\}N_1(0)$ \qquad (A3.4)

Substituting into the trial solution, we obtain:

$$N_2(t) = K\{exp(-\lambda_1 t) - exp(-\lambda_2 t)\}$$
$$= \{\lambda_1/(\lambda_2 - \lambda_1)\}N_1(0)\{exp(-\lambda_1 t) - exp(-\lambda_2 t)\}$$
$$= \{\lambda_1 N_1(0)exp(-\lambda_1 t)/(\lambda_2 - \lambda_1)\}\{1 - exp[-(\lambda_2 - \lambda_1)]t\}$$
$$= \{R_1(t)/(\lambda_2 - \lambda_1)\}\{1 - exp[-(\lambda_2 - \lambda_1)]t\}$$

Finally, as required we recover equation (3.10):
$$R_2(t) = \lambda_2 N_2(t) = R_1(t)\{\lambda_2/(\lambda_2 - \lambda_1)\}\{1 - exp[-(\lambda_2 - \lambda_1)]t\}$$

Exercises Chapter 3

Data on Atomic masses, radioisotopes, etc. can be found at [LBNL]:
http://ie.lbl.gov/toi2003/MassSearch.asp and
http://ie.lbl.gov/toi/nucSearch.asp.

1. 1.5g of copper is irradiated in a neutron source to produce radioactive ^{66}Cu via the reaction ^{65}Cu $+$ n $\rightarrow$ ^{66}Cu. The irradiation is stopped and the activity of the sample of copper is then measured with a Geiger counter. The initial measurement of 750 counts per minute drops to 210 in 9.4 minutes. Find the half-life and the decay constant of the radioactive copper. From the geometry of the source and counter, only 5% of all the disintegrations are counted. Calculate the initial number of radioactive atoms in the sample and the initial ratio of ^{66}Cu/ ^{65}Cu.

2. Polonium, $^{215}_{84}$Po , decays via alpha decay. Calculate the kinetic energy of the emitted alpha particle.

3. Plutonium–238, ^{238}Pu, emits an alpha particle of energy 5.10 MeV with a half-life of 90 years (2.84×10^9 s). 180 mg of this radioisotope are used to power a cardiac pacemaker. Calculate the initial power, in mW, supplied by the ^{238}Pu, assuming 100% efficiency of conversion of the decay energy to useable power.

4. ^{13}N decays decays via positron decay with a half life of 9.96 minutes. Write down the decay chain, and calculate the decay constant and the maximum kinetic energy of the positron.

5. $^{18}_{9}$F, used in PET scans, decays either by positron decay or by electron capture (EC). Write down the decay chains involved, and calculate the maximum energy of the emitted positrons for the β^+ decay and the Q value of the EC.

6. $^{38}_{17}$Cl, with an atomic mass of 37.968 010 u decays via β^- decay. Write down the decay chain, and calculate the maximum kinetic energy of the beta particle.

7. For the positron decay of a nucleus, calculate the minimum energy difference that must exist between parent and daughter in order for the decay to be energetically possible.

8. A parent nucleus, P, decays to a daughter, D, and an alpha particle. Show that the kinetic energy of the alpha particle is given by $Q \times m_D/(m_D + M_\alpha)$, where m_D is the mass of the daughter, m_α is the alpha mass, and Q is the disintegration energy of the decay.

9. A 1.5 kg tumour is being irradiated by an implanted radioisotope that delivers 12 Gy in 20 minutes. The gamma rays from the radioisotope have an energy of 0.5 MeV, all of which is deposited in the tumour. What is the activity of the radioisotope?

10. Radioactive ^{24}Na, which has a half life of 15 hours, is flown from Chalk River outside Ottawa to Princess Margaret Hospital in Toronto. It takes three hours for the transportation. If the activity must be 10 mCi when it is used in the hospital immediately on arrival, what activity must it have when it leaves Chalk River?

11. Dilution techniques measure the size of what is called the 'exchangeable pool'; this is that part of the material being measured that can exchange with the environment through ingestion and excretion. The exchangeable calcium in the body can be measured by dilution techniques. (Obviously the calcium deposited in the bones, for example, are not part of this exchangeable pool.) 3μCi of ^{47}Ca, (the radioactive isotope of calcium, which can be treated as if it had a very long life) is injected into the blood of a volunteer. 24 hours later it was found that 14% of the original 3μCi had been voided in the urine. A sample of blood taken at that time was found to contain 620 μg of stable calcium, and 0.21 nCi of ^{47}Ca. Calculate the mass of exchangeable calcium in the volunteer's body.

12. 1.20×10^5 Bq of ^{42}K which has a nuclear half-life of 12.5 hours were injected into the body. 24.0 hours later some blood was taken; this sample had 0.150 g of stable potassium and 29.0 Bq of ^{42}K in it. Neglecting biological excretion (the biological half life is long), and assuming the potassium had fully mixed throughout the body, which of the following values is closest to the amount of stable potassium in the body in grams?

13. A body contains 120 g of exchangeable potassium. 3.2×10^4 Bq of ^{42}K (half-life of 12.5 hours) was injected into the body. After 24 hours, a sample of serum was found to contain 200 mg of stable potassium. A) Calculate the activity of ^{42}K in the sample. B) What will be the

activity of ^{42}K in another 200 mg sample taken 12 hours later? Neglect biological excretion, but take nuclear decay into account.

14. Uranium, along with other heavy elements on earth, was created by nuclear fusion reactions in supernovae in the early universe. 99.28% of the naturally occurring Uranium on earth consists of the ^{238}U isotope; the isotope ^{235}U accounts for the remaining 0.72%. Tables showing the decay chains for ^{235}U (the so-called Actinium series) and also for ^{238}U (the Radium series) are available at http://en.wikipedia.org/wiki/Decay_chain. You will need the numbers in these tables to answer some parts of this question. You will also need to know that our bodies contain approximately 2.0×10^{-5} g of Uranium, a mass that is kept fairly constant over time by a balance between ingestion (mainly breathing) and excretion. A) Making the reasonable assumption that ^{238}U and ^{235}U were originally created in equal numbers, how long ago was the supernova explosion that created them? B) Consider the first decay of ^{238}U to ^{238}U. Verify the quoted value of alpha particle energy from this decay.

Chapter 4

Radiation Therapy

4.1 Introduction

Previous chapters have described the use of radiation as a diagnostic tool. However, in its passage through matter, radiation causes ionization that can damage the cells through which it passes. At the low levels of radiation used in diagnosis, the body can usually repair this damage; at higher levels, illness can result. However, since radiation preferentially destroys cancer cells, which grow at a faster rate and repair themselves less well than do healthy cells, radiation has become a powerful tool in the treatment of cancer.

In radiation therapy, the intent is to sterilize or kill the tumour cells while causing minimal damage to adjacent healthy tissues. Many different sources of radiation are used. There are two main methods of delivery of the radiation: **teletherapy**, where the radiation from X-rays, radioisotopes, or external particle beams (electrons, protons, neutrons, or heavy ions) is directed into the body from the outside, and **brachytherapy** (the root comes from Greek, meaning short distance), in which a radioisotope is placed on or inside the body, close to the site of the cancerous tissue.

Section 4.2 discusses the mechanisms whereby biological cells are damaged by radiation. Section 4.3 shows some basic methods used to calculate the required therapeutic doses from external radiation. Section 4.4 discusses beams of radiation from accelerators and from radioisotopes. Brachytherapy, its uses, and related dose calculations appear in section 4.5.

54

Chapter 5 provides a short overview of the unavoidably deleterious effects of radiation on healthy cells and the contribution of both diagnostic and therapeutic radiation to the environmental radiation background.

4.2 The interaction of radiation with biological cells

X-rays and the emissions from radioactive isotopes (alpha, beta, or gamma rays) ionize the cells through which they pass. (Ionization is the process whereby electrons are removed from an atom or molecule to form an ion or a charged molecule). Alpha particles (nuclei of He atoms, see §3.2.2), by virtue of their greater mass and double charge, produce a greater **density** of ionization than beta particles (electrons or positrons). As a consequence, alpha particles are much more weakly penetrating than beta particles, which in turn are less penetrating than X-ray or gamma ray photons at a given energy. Alpha particles of typical energy are easily stopped by, e.g. a sheet of paper, or human skin; however if they are ingested, their high ionizing ability causes cellular damage to the lungs, which is the main reason smoking causes cancer (see §5.1.2). Photons of X- or gamma rays are called *indirectly ionizing* radiation since they do not directly cause biological damage, but produce energetic electrons via the processes described in §2.3 (the photoelectric effect, Compton scattering, and pair production) that, in turn, cause damage through multiple ionizations.

The **biological effects** are proportional to the amount of ionization caused by the radiation, which, in turn, is proportional to the amount of energy deposited. Since an energy deposition of around 30 eV produces one ion pair (§2.3.5), X-ray or gamma radiation produces thousands of ion pairs.

Experimental studies show that ionizing radiation causes damage to cells of the body by breaking one or more DNA strands, either directly or by the interaction of the O and OH free radicals produced by the irradiation of the H_2O molecules. Either the sugar phosphate backbone of the DNA or the base-pair sequence of the DNA can be broken. The living cell has repair enzymes to repair breaks in one strand. However, if both strands of the DNA backbone are broken, either by the same interaction or by two nearly simultaneous interactions, the damage may be irreparable. In that case, the cell may become sterile or die. Ironically, while this effect is used to kill cancer cells, it can have a deleterious long–term effect in healthy cells, since if a healthy cell survives in a defective form and continues to divide and reproduce, cancer may be the result.

Figure 4.1 indicates schematically the radiation sterilization of cancer cells, based on a model that assumes that the breaking of both DNA strands leads to certain cell sterilization. While cancer cells are more susceptible to damage than healthy cells, it is only above a dosage of 2Gy that the total survival is very small. This 'linear–quadratic' model is used in calculations of radiation damage.

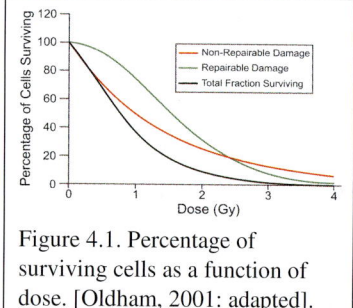

Figure 4.1. Percentage of surviving cells as a function of dose. [Oldham, 2001: adapted].

4.3   The absorbed dose from external radiation

The biological effects of radiation depend on three factors: the energy deposited per unit mass, the distribution of the energy deposition, and the susceptibility to damage of the traversed tissue. This section explains how these effects are calculated for beams of radiation.

4.3.1   *Absorbed dose, and energy fluence*

Units of **exposure** (the roentgen (R)) and the **absorbed dose** (the rad and the Gray (Gy)), have been defined in §2.3.5. In order to estimate the effects of a beam of X-rays or gamma rays on biological matter, we need to know how to calculate the absorbed dose (in Gy) from knowledge of the energy carried by the beam, or, in the case of X-rays, of the exposure (in R). The fundamental quantity is the energy carried in the beam of radiation. We will start by considering how this energy is influenced by its interaction with material through which it passes.

Consider a beam of X-rays or γ-rays that carries a certain amount, Ψ, of energy per unit of cross-sectional area: this is called the energy fluence, measured in J m^{-2}. (Note the similarity to the photon fluence, Φ, defined in §2.4). If the beam is monoenergetic, Ψ is simply equal to the energy of each photon, e_γ, multiplied by Φ, the number of photons per cross-sectional area.

In §2.4, where we were interested in the ability of a beam of X-ray photons to form a sharp image, we calculated the loss of photons scattered *out* of a beam by a small slab of material, of thickness dx at a depth x in the material. Here we are interested in the energy deposited *in* such a small slab by the interaction of the photons with the atoms of the material. The calculation proceeds in exactly the same manner, with dE replacing dN, and Ψ replacing Φ. In this case, the linear attenuation coefficient μ, is replaced by a new constant, the **energy absorption coefficient,** μ_{en}, which, similarly to μ, depends on the type and energy of radiation and the properties of the material, including its density

In analogy with equation (2.2), the reduction in energy fluence at depth x is given by: $d\Psi = -\mu_{en}\Psi(x)dx$ (4.1)

The energy deposited in the slab is this reduction in the energy fluence multiplied by the surface area : $dE = |d\Psi| = \mu_{en}\Psi(x)Sdx$ (4.2)

In analogy with equation (2.3), the diminution of energy of the beam of photons with depth in the material, at a depth x, is given by:

$$\Psi(x) = \Psi(0)exp(-\mu_{en}x)$$ (4.3)

Note that Ψ is written as $\Psi(x)$ to remind you of the dependence of Ψ on the position in the material, x.

The energy fluence can now be related to the absorbed dose, which is the quantity of most interest in medical applications. The absorbed dose at a depth x, $D_m(x)$, is defined as the energy absorbed by the slice of the material (designated by the suffix m) at the depth x (thickness dx and surface area S) divided by the mass, ΔM, contained in the slice; this mass is equal to the density, ρ, times the volume: $\Delta M = \rho S\, dx$. Thus:

$$D_m(x) = (dE/\Delta M) = (\mu_{en}\Psi(x)S\,dx/\rho S\,dx) = (\mu_{en}/\rho)_m\Psi(x) \quad (4.4)$$

The quantity $(\mu_{en}/\rho)_m$ is called the **mass energy absorption coefficient.** Some values of $(\mu_{en}/\rho)_m$ are given in Appendix A4.1. Figure 4.2 shows the behaviour of μ_{en} with photon energy for three biological materials of interest. As expected, bone absorbs energy much more than muscle or tissue.

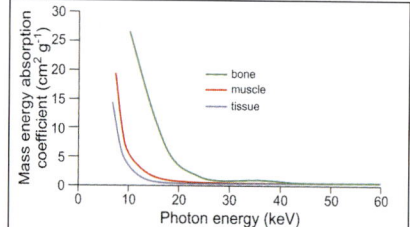

Figure 4.2. Mass energy absorption coefficient as a function of photon energy [plotted from data in Hubbell and Seltzer, 2004, NIST; see Table A4.1].

The energy absorption coefficients for air are approximately 10^{-3} smaller than those for biological tissue, due to the much lower density of air. Therefore, useful absorption of X-rays can be achieved in air only by increasing the distance from the source (giving an approximately $1/r^2$ reduction) or, most commonly, by inserting a thickness of heavy metal.

We are now in a position to calculate the **average absorbed dose, $\overline{D_m}$**. This is the energy per unit mass imparted to sample of biological material by radiation. Let a mass M, of density ρ, be subject to an exposure of a beam of X-rays, of cross-sectional area S, and exposure X Roentgens. Let the thickness of the material be L, and let the air-tissue interface be at $x = 0$. First, we calculate the energy fluence of the beam. From §2.3.5 we have $D_{air} = 8.69 \times 10^{-3}X$ Gy. Combining with equation (4.4) for air yields the energy fluence in air, at the surface:

$$\Psi_{air} = \Psi_m(0) = D_{air}/(\mu_{en}/\rho)_{air} = 8.69 \times 10^{-3}X/(\mu_{en}/\rho)_{air} \quad (4.5)$$

(If X is in Roentgens and $(\mu_{en}/\rho)_{air}$ in m^2 kg^{-1}, Ψ is given in J m^{-2}.)

After traversing the thickness L, the energy fluence of the X-rays is given by Equation (4.3) with $x = L$, so the total energy, E_{tot}, deposited in the mass is equal to the product of the surface area S, multiplied by the difference between the entering and the exiting energy fluence:

$$E_{tot} = S\{\Psi_{air} - \Psi_m(L)\} = S\Psi_{air}\{1 - exp(-\mu_{en}L)\} \quad (4.6)$$

Division by $M = \rho SL$ yields the average absorbed dose of the material that is irradiated in terms of the incident energy fluence:

$$\overline{D_m} = E_{tot}/M = \Psi_{air}\{1 - exp(-\mu_{en}L)\}/\rho L \qquad (4.7)$$

Combining equations (4.5) and (4.7), we obtain the required expression for X-rays in terms of the exposure, X:

$$\overline{D_m} = 8.69 \times 10^{-3}X\{1 - exp(-\mu_{en}L)\}/\{\rho L(\mu_{en}/\rho)_{air}\} \qquad (4.8)$$

4.3.2 Radiation weighting factors and equivalent dose

With 8000 times the mass and double the charge, a slow alpha produces a much greater density of ionization than an electron that deposits the same total energy; both alpha and electron produce more density of ionization than a photon. Thus even though a photon may deposit the same total energy as an alpha, it does less biological damage.

These effects are taken into account by the "radiation weighting factors", $W_R{}^*$, published by the International Commission on Radiological Protection [ICRP, 2007]. The average absorbed dose (in Gy or rad) is converted to an '**equivalent dose**' by multiplying by W_R. The unit of the equivalent dose is the **Sievert (Sv)**, derived directly from the Gray, named after Swedish scientist Rolf Sievert (1896 –1960). An older unit, still used, is the **rem** (roentgen equivalent mammal).

Table 4.1. ICRP Radiation weighting factors for different types of radiation.			
Radiation Type	W_R	**Radiation Type**	W_R
Photons*, electrons	1	α-particles	20
Protons	2	neutrons	function of energy
*$W_R = 1$ holds over a wide range of X-ray energies.			

Thus, the **equivalent dose** in Sv is $H = W_R \times$ (absorbed dose in Grays). The rem equivalent is given by rem $= W_R \times$ (absorbed dose in rads). 1 Sv = 100 rem; 1 mrem = 10 μSv.

---

* **Relative biological effectiveness – RBE,** and '**Quality Factor**', denoted by **Q** or **QF**, are other terms in common use, with slightly different meanings and values from W_R.

Table 4.1 gives some typical values of the radiation weighting factor for different types of radiation traversing "average" biological tissue. The larger values, for protons, alpha particles, and neutrons, correspond to the fact that these particles do more biological damage than X- or gamma rays in a given tissue.

4.3.3 *Tissue weighting factors and effective dose*

As we will see in Chapter 5, the calculation of radiation risk is complicated. For the same density of ionization, different tissues are more susceptible to damage than others and they also have different susceptibilities to different types of risk, such as fatal or non-fatal cancers, hereditary risks, etc. In order to compare the risks arising from a given diagnostic or therapeutic dose (usually directed to only a fraction of the body) to the environmental radiation background (to which the whole body is exposed) a calculation is done to normalize the dose to a 'whole body' dose that would carry the same risks.

The ICRP [2007] also publishes 'tissue weighting factors', W_T, to take these effects and the size of the different organs into account for radiation protection purposes. The values, averaged over gender and age, are shown in Table 4.2. Higher values of W_T correspond to greater susceptibility to risk. The equivalent dose for each irradiated organ, H_T, is multiplied by W_T and summed over all irradiated organs to obtain what is called the **effective dose**: $H_{eff} = \sum W_T H_T$.

Table 4.2. ICRP Tissue weighting factors for different biological tissues.

Organ or Tissue	W_T	ΣW_T
Colon, Lung, Breast, Stomach, Bone Marrow	0.12	0.60
Gonads	0.08	0.08
Bladder, Oesophagus, Thyroid, Liver	0.04	0.16
Bone, Skin, Salivary Glands, Brain	0.01	0.04
Other Organs (grouped)	0.12	0.12

The effective dose (H_{eff}) is intended to allow direct comparison of the biological risks (cancer and hereditary effects, for example) caused by different radiation exposures to different parts of the body. Though the calculation of the tissue weighting factors is difficult, the idea is simple. Since the type of radiation and the type of the tissue irradiated affect the level of risk, both radiation and tissue weighting factors are used. The weighting factor, W_T, is defined in such a way that the sum of the weighting factors taken over all organs of the body is equal to 1: $\Sigma\ W_T = 1$. If only a fraction of an organ is irradiated, the result for that organ must be multiplied by that fraction.

For example, an equivalent dose of 1 Sv to the gonads ($W_T = 0.08$) is expected to produce as great a risk of fatal cancer as a whole body dose of 0.08 Sv. An equivalent dose of 3Sv to 50% of the colon is expected to produce as great a risk of fatal cancer as a whole body dose of $0.5 \times 3 \times 0.12 = 0.18$ Sv.

The calculation of the risks from different effective doses is discussed in Chapter 5.

4.4 Teletherapy

With increasingly sophisticated technology, and after many clinical trials, the calculation of the dose required for a therapeutic procedure has become very precise. Large computer programmes, often using Monte Carlo techniques, are employed in the calculations. Usually a combination of extensive pre-operation scans—ultrasound, CT, and MRI—are used to provide detailed information to the computer driven procedures. Many sophisticated filters are then programmed to ensure that the dose to the diseased area is maximized, while the dose to any intervening tissue is minimized.

4.4.1 *Accelerator beams*

One beautiful example of an electron beam treatment is shown in Figure 4.3, which shows the results of a calculation designed to avoid dose to the lungs [Rogers,2002]; a computer-generated Monte Carlo 'dose engine' generates the isodose lines (the coloured lines along each of which the dose is constant).

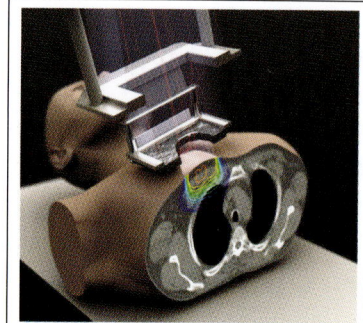

Figure 4.3.Calculated isodose lines for an electron beam treatment. [Courtesy Canadian Association of Physicists].

X-rays, gamma rays, and electrons are used for radiation therapy. However, photons penetrate further in tissue than electrons (see §2.4, Figure 2.6). Thus X-rays or gamma rays in focused beams can produce electrons at depth without depositing an excessive dose to the intervening healthy tissue. X-ray energies of several MeV are used for this application. Electron beams, or 'softer' X-rays (10 to 100keV) at high exposures are most useful for treating tumours that are close to the surface of the body.

Beams of other particles have also been used, though these require more complicated and expensive machinery. Protons, or nuclei of helium and more massive charged particles have particularly useful properties; they can be precisely focused and, since the rate of energy loss increases rapidly with decreasing particle speed, they give rise to a sharp rise in the ionization near the end of their range. Thus, for such particles, the maximum ionization volume can be precisely defined. There are around 40 proton therapy facilities worldwide, about a dozen in the US. In Canada, the only proton radiation therapy facility is located at the Tri-Universities Meson Factory accelerator (TRIUMF) in Vancouver. It is mainly used for treating eye tumours, where very precise irradiation is required.

Since healthy cells recover faster than cancer cells, a large number of smaller doses delivered over a period is more effective than one large dose delivered at one time. This is called the **fractionation** of the dose. When treating a tumour with a fractionated dose, the beam is directed at the tumour from different directions at each exposure to maximize the dose to the tumour while minimizing the exposure to the surrounding healthy tissue. This is called **rotation therapy**. A total dose of 50 to 60 Gy, spread over 25 to 30 daily sessions is typical.

4.4.2 *Beams from radioisotopic radiation*

High activity radioisotopes can provide intense high energy gamma ray beams that are directed at diseased areas in a similar way to X-ray or electron beams.

The most common radioisotope used in teletherapy is ^{60}Co, made by neutron activation of ^{59}Co, via ^{59}Co + n $\rightarrow$ ^{60}Co, using neutrons from a nuclear reactor. Much of the world's supply of this isotope used to be made at Chalk River Laboratories in Ontario (Atomic Energy of Canada Limited, AECL), and the first Cobalt therapy unit was brought into operation in London, Ontario, by Harold Johns in 1951.

The advantages of Cobalt 'bomb' units include their compact size, the high energy and intensity of the gamma rays, and their relatively low cost; many clinicians prefer them for treatment of head, neck and breast cancer. However they cannot be switched off, so require great care in shielding medical personnel, and the source must be replaced every 3 to 5 years (the half life is 5.26 years). Although they are gradually being replaced by the more expensive and bulky linear accelerators, Cobalt units are still used in regions of the world where there is not easy access to the technological expertise required for the former.

The effective energy of radiation from an X-ray machine is approximately one third of the maximum energy of the X-rays. ^{60}Co produces gamma rays of energies 1.17 and 1.33 Mev, with an average energy of about 1.25 MeV. Thus, approximately, the radiation from the

Cobalt machine is as penetrating as that from a 3 MV X-ray machine. A typical ^{60}Co radiation unit uses a source of up to 100TBq (terra Bq = 10^{12} Bq).

^{137}Cs, another previously popular isotope obtained from the fission products of a nuclear reactor, is no longer being produced for teletherapy; the gamma rays (662 keV) do not penetrate sufficiently to treat deep seated tumours, and the specific activity (the activity per gram) of ^{137}Cs is not so high as that of ^{60}Co.

4.5 Brachytherapy

In brachytherapy, the radioisotopes are introduced into the body by ingestion or implantation, where the emitted radiation attacks a localized tumour.

Brachytherapy has a long history; radium was used to treat cancer within five years of its discovery, by placing tubes or needles of the radioisotope in or on tumours. In Quebec and French speaking countries it is called Curietherapie in honour of Marie Curie, the only person to receive Nobel prizes in both Physics (in 1903, shared with her husband, Pierre Curie, and Henri Becquerel) and Chemistry (in 1911). Often beta emitters are preferred, due to the short range of beta particles in tissue, which ensures that the damage caused by the radiation is limited to the organ being targeted. Therapeutic radioisotopes should also have short half lives to reduce unwanted long-term effects.

Brachytherapy can be either temporary, in which the radioisotope is withdrawn after a fixed time, or permanent, in which lower activity radioactive seeds are left in the patient's body.

The thyroid gland accumulates Iodine. If the gland becomes cancerous, a dose of radioactive iodine, administered orally, can attack the cancerous cells. Since iodine does not accumulate in other parts of the body, the damage to healthy organs is limited. Several iodine isotopes are in common use (e.g. ^{123}I, ^{131}I) depending on the therapeutic requirements.

Another cancer often treated with brachytherapy is prostate cancer, the second leading cause of cancer death of males after lung cancer; approximately 1 in 8 males will contract prostate cancer. Titanium catheters containing a suitable radioisotope (e.g. ^{125}I with a half life, $T_{1/2}$, of 60 days, or ^{103}Pd with $T_{1/2} = 17$ days) are prepared and inserted through the perineum, guided by ultrasound images in real time. For aggressive cancers, temporary implants of high activity are used; for lower risk cancers, permanent low activity implants are used. The latter method of treatment has been found to be as efficacious as chemotherapy or surgery, with a much shorter hospital stay (usually one day) and less severe side effects.

Temporary brachytherapy is common in the treatment of breast cancer, one of the leading cause of death in women (about 1 woman in 9 will develop breast cancer). Commonly used isotopes are ^{137}Cs (half-life of 30 years and a gamma energy of 0.662 Mev), ^{192}Ir (half-life of 74 days and a gamma energy of 0.4 Mev) and ^{103}Pd (half-life of 17 days and a gamma energy of 21 kev). ^{103}Pd has been used to treat scar tissue after a breast operation; it releases continuous low doses of radiation, replacing the conventional treatment of from two to seven months of daily radiation (of the order of 50 Gy delivered in 25 fractions) that requires regular hospital visits and often produces unpleasant side effects.

The following sections presents some basic concepts used to calculate the required dose when radioisotopes are administered internally.

4.5.1 *The dose from brachytherapy*

Consider the ingestion or implantation of a radioisotope into an organ of the body for therapeutic purposes. We will calculate the total absorbed dose expected over the lifetime of the radioactivity in the body.

First consider the case where the radioisotope and its decay products remain in the body for a time that is essentially infinite compared to the short half-life of the radioisotope. Then the total absorbed dose, denoted by D_m $(0{\rightarrow}\infty)$ is just equal to the total number of radioactive nuclei introduced (assuming almost all have decayed), multiplied by the energy deposited by the products of each nuclear disintegration, e_n, divided by the mass of the organ in which the radiation is deposited.

That is, $D_m(0 \rightarrow \infty) = e_n N(0)/m$ $\hspace{3cm}$ (4.9)

where $N(0)$ is the number of radioactive isotopes introduced into the organ at time $t = 0$, and m is the mass of the organ. $N(0)$ is given by the usual formula: $N(0) = R(0)/\lambda$ $\hspace{3cm}$ (4.10)

where λ is the disintegration constant. Thus the dose is:

$$D_m(0 \rightarrow \infty) = e_n R(0)/(m\lambda) \hspace{2cm} (4.11)$$

This can also be written as:

$$D_m(0 \rightarrow \infty) = \dot{D}_m(0)/\lambda \hspace{2cm} (4.12)$$

where the initial dose rate is written as:

$$\dot{D}_m(0) = [dD_m(t)/dt]_{t=0} = e_n R(0)/m \hspace{2cm} (4.13)$$

As expected, the total dose received is proportional to the activity of the radioisotope, its energy and its half-life, and inversely proportional to the mass of the organ.

In many cases the implanted radioisotope is removed after a certain time, in order to ensure the delivery of the correct dose required for treatment. This is a special case of the one calculated in the next section; see equation (4.23).

4.5.2 *Biological excretion*

Of course, the natural decay of the radioisotope may not be the only mechanism whereby the radioactive nuclei disappear from the body or from an organ into which they have been introduced. Sometimes the introduced radioisotope is removed after some time by a medical procedure. Also, in all realistic cases, the biological processes of excretion must also be taken into account. The situation can be complicated—for example, if the radioisotope is stored in the gut or bladder. However, for simplicity of calculation, it is usual to assume that the disappearance of an isotope from a particular organ caused by biological processes follows an exponential law.

Let us generalize the result in §4.5.1 to consider the case in which a radioisotope is introduced into an organ at time $t = 0$ and is removed after a time $t = T$, during which biological excretion operates. Assume that all of the energy from those nuclei that decay in the organ contribute to the absorbed dose in the organ; let each decay deposit energy e_n. Use the suffixes 'n' to denote nuclear decay, and 'be' to denote biological excretion. Remember that, for any number, $N(t)$, of radioactive nuclei at time t, the activity—the number of *radioactive* decays per second—is given by $\lambda_n N(t)$.

At time t, the number of nuclei that decay in the organ in an interval of time dt due to nuclear decay can thus be written:

$$dN_n(t) = -\lambda_n N(t)dt \tag{4.14}$$

where λ_n is the usual nuclear decay constant.

Similarly, the number of nuclei that disappear from the organ in time dt due to biological excretion can be written (using the exponential assumption) : $dN_{be}(t) = -\lambda_{be} N(t)dt$ (4.15)

where λ_{be} is called the **biological decay constant**. These are the fundamental equations.

First we use them to derive a couple of useful results. The total number of radioactive nuclei that disappear from the body in dt, at time t is the sum of the numbers above:

$$dN(t) = dN_n(t) + dN_{be}(t) = -\lambda_n N(t)dt - \lambda_{be} N(t)dt$$
$$= -(\lambda_n + \lambda_{be}) N(t)dt = -\lambda_{eff} N(t)dt \tag{4.16}$$

where $\lambda_{eff} = \lambda_n + \lambda_{be}$, is called the **effective decay constant**.

The solution of equation (4.16) (see §2.4 and §3.2) yields an expression for the number of nuclei remaining in the organ at time t:

$$N(t) = N(0)exp(-\lambda_{eff}t) \tag{4.17}$$

N(0) is the number of radioactive isotopes introduced at time $t = 0$.

Note that only a fraction of the total number, *N(t)*, of nuclei in the organ at time t, actually contribute to the absorbed dose in the organ by decaying; the rest are excreted before they deposit any energy. From (4.14) and (4.15) this fraction is:

$$f = dN_n(t)/dN(t) = \lambda_n/\lambda_{eff}, \tag{4.18}$$

Since *f* is independent of the time, it must be the same fraction for any time interval—including the one from $t = 0$ to $t = T$.

We now proceed to calculate the dose delivered to the organ. The number of nuclei disappearing from the organ—by decay or biological excretion—is the difference between the number of nuclei at time $t = 0$ and the number remaining at $t = T$. Using equation (4.16), we obtain:

$$N(0) - N(T) = N(0)\{1 - exp(-\lambda_{eff}T)\} \tag{4.19}$$

The fraction of these nuclei that deposited their energy in the organ is *f*. Thus the energy deposited by these decays from $t = 0$ to $t = T$ is:

$$E_m(0 \rightarrow T) = e_n f\{N(0) - N(T)\} = e_n f N(0)\{1 - exp(-\lambda_{eff}T)\}$$
$$= e_n(\lambda_n/\lambda_{eff})N(0)\{1 - exp(-\lambda_{eff}T)\} \tag{4.20}$$

where we have substituted for *f* from equation (4.18).
Substituting $N(0) = R(0)/\lambda$ from equation (5.2), we obtain:

$$E_m(0 \rightarrow T) = e_n R(0)\{1 - exp(-\lambda_{eff}T)\}/\lambda_{eff} \tag{4.21}$$

Thus, finally, the dose delivered in this time interval *T* is:

$$D_m(0 \rightarrow T) = E_m(0 \rightarrow T)/m = e_n R(0)\{1 - exp(-\lambda_{eff}T)\}/(m\lambda_{eff}) \tag{4.22}$$

If biological excretion can be ignored, λ_{eff} becomes equal to λ_n:

$$D_m(0 \rightarrow T) = e_n R(0)\{1 - exp(-\lambda_n T)\}/(m\lambda_n) \qquad (4.23)$$

In this case, if the radioisotope is not removed from the organ, the dose to infinity is obtained by allowing $T \rightarrow \infty$, to obtain $D_m(0 \rightarrow \infty) = e_n R(0)/(m\lambda_n)$, which is just equation (4.11).

These calculations assume that the biological 'uptake' time is instantaneous. Appendix A4.2 gives the result when the time for the organ to absorb the radioisotope cannot be neglected.

A4.1 *Table of mass energy absorption coefficients*

Table A4.1. Mass energy absorption coefficients $(\mu_{en}/\rho)_m$ $(cm^2\,g^{-1})$ [Hubbell and Seltzer, 2004, NIST]

Energy (keV)	Air	Tissue	Muscle	Bone
10	4.742	2.935	4.964	26.80
20	0.539	0.325	0.564	3.601
40	0.068	0.046	0.072	0.451
60	0.030	0.026	0.033	0.140
80	0.024	0.024	0.026	0.069
100	0.023	0.024	0.025	0.046
200	0.027	0.030	0.029	0.030
500	0.030	0.033	0.033	0.031
1000	0.028	0.031	0.031	0.029
2000	0.024	0.026	0.026	0.024
5000	0.017	0.019	0.019	0.019
10000	0.015	0.015	0.015	0.016
20000	0.013	0.013	0.014	0.016
Density (g.cm^{-3})	0.001293	0.95	1.00	1.85

A4.2 *Brachytherapy dose with biological uptake*

If the time required for an organ to absorb an ingested radioisotope is comparable to the nuclear or biological half-life, the calculation given in §4.5.2 overestimates the dose. The calculation is complicated, depending on the precise assumptions made. If the uptake can be assumed to be exponential with a half-life of T_{bu} , the total absorbed dose in the organ is given by:

$$D_m(0 \to \infty) = \left(1.44 \times e_n R(0) T_{eff}/m\right)\left(1 - T_{up}/T_{eff}\right) \qquad \text{(A4.1)}$$

where $1/T_{up} = 1/T_{eff} - 1/T_{bu}$.

Exercises Chapter 4

1. The transport of radioactive samples must follow strict guidelines to ensure that the exposure is very low. A 100 mCi sample of ^{212}Pb is placed at the centre of a cubical lead lined box, each side of which has a length of 30cm. Calculate the thickness of lead required to reduce the exposure level at the surface of the box to 10 mR per day at most. The half-life of ^{212}Pb is 11 hours, and each gamma disintegration yields 0.24 MeV; the mass energy absorption coefficient for lead at this energy is 0.48 cm^2g^{-1}, and its density is 11.4 g cm^{-3}. (Make the simplest reasonable assumption you can, and provide a full explanation).

2. A diagnostic X-ray machine operates at 180 kV peak voltage. It is customary to assume that the average energy of X-rays from such a machine is one third the maximum possible. A) find the energy and wavelength of the most energetic X-rays. B) One metre away from the machine the exposure rate is measured to be 1R per minute. The recommended maximum dose for the technician is 10^{-4} R per minute for a 40 hour week. How thick must the lead shielding be to ensure that the technician's exposure remains within acceptable limits? (for lead, which has a density of 11.4 g.cm^{-3}, the mass energy absorption coefficient at 60 keV is 3.8 cm^2g^{-1}).

3. The whole body of a person of mass 70 kg and average thickness from back to front of 25 cm is exposed for 5 seconds to a uniform fluence rate of 2 MeV gamma rays of 3×10^6 photons per cm^2 per

second. Assume the average density of the tissue and bone traversed is 1.2 g.cm^{-3}. A) Calculate the total number of photons entering the person. B) If 60% of the energy of the gamma rays is deposited in the person's body, calculate the absorbed dose rate. C) calculate the equivalent dose and D) calculate the effective dose.

4. A 95 kg miner inhales 500 nCi of ^{60}Co. ^{60}Co decays to gamma rays, of average energy 0.75 Mev, with a nuclear half-life of 5 years; the biological half-life in the lungs is 20 days. Calculate A) the absorbed dose, B) the equivalent dose, and C) the effective dose the miner receives from this exposure.

5. A given radioisotope, with a half-life $T_{1/2}$ is implanted in the prostate, of mass m, with an initial activity of R(0). It remains there for a time T, and is then surgically removed. Ignoring biological excretion, derive an expression for the total dose in terms of $T_{1/2}$ and T.

6. A 65 kg worker, whose cross sectional area can be assumed to be 1.5 m^2, is exposed to a 35 mCi ^{60}Co source, 3 m. distant, for 4 hours per day. ^{60}Co emits two gamma rays at each decay, of energies 1.33 MeV and 1.17 MeV. If 50% of the gamma rays deposit their energy in the worker's body, what is the whole-body dose received?

7. A 60 kg patient had 12×10^6 Bq of a radioisotope injected into her body. The radioisotope has a nuclear half-life of 20 hours and a biological half-life in the body of 2.5 days. The radioisotope yields 1.5 MeV per disintegration as beta and gamma rays, of which 40% is absorbed in the body. What is the dose received by the patient?

8. 3.0×10^6 Bq of ^{24}Na were injected into a person's body. 15 hours later a sample of body fluid was measured to have 0.1 g of stable sodium and 1.5×10^3 Bq of ^{24}Na. 30 hours later another sample was measured to have 0.1 g of stable sodium and 0.5×10^3 Bq of ^{24}Na.. Calculate the effective half life and the total mass of exchangeable sodium in the body.

9. A hospital buys a 3000 Ci source of ^{60}Co that emits gamma rays of effective energy 1.25 MeV. An unfortunate technician, standing at 2.0 m from the source, receives an exposure to the upper 50% of his body, of area 35×35 cm^2, for 0.75 s.

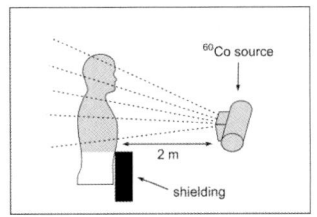

Assume all of the gamma rays in the beam traverse his body, that he is 25 cm thick and that his density is 1.1 g.cm^{-3}. The mass energy absorption coefficient for tissue at this energy is 0.026 cm^{2}g^{-1}. A) What is the energy fluence rate at the front surface of his body (ignore absorption in the air). B) What is the total energy that enters his body from these gamma rays during this exposure? C) What is the total gamma ray energy that leaves his body? D) What is the absorbed dose he receives from the energy deposited by these gamma rays? E) What is the effective dose he receives from this exposure? State clearly any assumptions you make about the values of the weighting factors W_R and W_T.

10. ^{32}P has a nuclear half-life of 14.3 days and an effective half-life in the body of 13.5 days. Calculate the biological half life.

11. Our bodies contain approximately 2.0×10^{-5} g of Uranium, a mass that is kept fairly constant over time by a balance between ingestion (mainly breathing) and excretion. Use the information in question 14 in the exercises of Chapter 3. A) Estimate the absorbed dose that you receive annually from the ^{238}U in your body. State clearly any assumptions you require. Explain why you need consider only the first decay in the decay chain that involves alpha emission. B) Now estimate the absorbed dose that you receive annually from the ^{235}U in your body. You will notice that ^{239}P is the parent of ^{235}U. C) Make a numerical argument for ignoring this isotope in your calculation; remember that we are considering only natural sources for the radioisotopes. (In fact, there is some concern about Plutonium contamination from the atmospheric nuclear weapon tests, but that is another story!). Again, explain why you need consider only the first decay in the ^{235}U chain that involves alpha emission. C) Finally, estimate the total absorbed dose, the equivalent dose, and the effective dose that you receive annually from the Uranium in your body.

Chapter 5

Radiation and the Environment

5.1 Environmental exposure to radiation

The background radiation to which we are continuously exposed provides a useful baseline to evaluate the potential risks of the diagnostic and therapeutic radiation we have discussed so far. This background includes natural and man-made radiation, from cosmic rays, from the air we breathe, from the ground, from buildings, from our food, even from our own and others' bodies.

Many national and international organizations provide values of the background radiation, both natural and man-made. Precise values vary from country to country, from region to region, from individual and individual, and from year to year. The figures in this chapter are intended to give an overall understanding, with, however, only representational numbers.

The global average effective dose per person from all sources is about 3 mSv. The greatest contributor to natural sources is Radon, the greatest contributor to man-made radiation, at least in the developed countries, are medical procedures. In N. America, both sources of radiation are greater than in many other parts of the world leading to a total of about 6 mSv per year per person.

5.1.1 The natural radiation background

Globally, about 80% of the total radiationwe receive comes from natural sources. The worldwide average effective dose is 2.4 mSv per year, with huge variation from place to place—from 1 mSv/yr to over 700 mSv/yr!

An average North American resident receives an annual dose from natural radiation of around 3 mSv; people living at high altitudes or on the Canadian Shield, for example, receive much higher doses from the increased cosmic ray background or the terrestrial radiation respectively.

Figure 5.1 shows the various contributors to this background. The gas Radon is the most important. Chemically inert, with a half-life of 3.8 days, Radon is one of the by-products in the decay chain of ^{238}U and ^{232}Th, which are present in most rocks and soil at levels of several parts per million, and ten times that in some rocks such as granite and

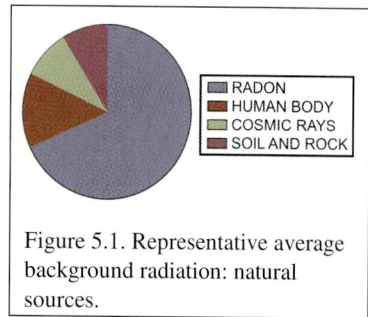

Figure 5.1. Representative average background radiation: natural sources.

sedimentary rocks. Both of these isotopes have half-lives of several million years, and both decay chains produce Radium, which decays into isotopes of Radon, mainly $^{222}_{86}$Rn . (The radioactivity of this and similar decay chains generate heat in the earth's crust, which maintains the temperature of the earth necessary to sustain life on earth). The Radon decay chain produces a number of radioactive heavy metals, mainly polonium, bismuth, and lead. These nuclei, being charged, can attach themselves to particles of dust which are then inhaled; the overall half-life of the Radon daughters is 22 years. In the open air, depending on location, the radioactive level is several tenths of a pCi per litre of air. Inside the underground basements of buildings, Radon can accumulate tenfold. The US Environmental Protection Agency recommends that domestic basement levels should not exceed 4 pCi per litre [EPA]. (A useful conversion factor is 2.7 pCi/g = 100 Bq/kg). After smoking, it is the second leading cause of lung cancer.

The next main contributors to the natural background are our own bodies. The food we eat has many radioactive isotopes absorbed from the ground or from the air. Table 5.1 shows the amounts of radio isotopes in the body of an "average" 70 kilogram adult. ^{40}K and ^{14}C are the main contributors by far.

Table 5.1. Some of the radioisotopes in an 'average' human body [ICRP, 1975].

Nuclide	Activity (pCi/70kg)
Uranium–238	30
Potassium– 40	120,000
Radium–226	30
Carbon–14	100,000
Polonium	1,000

The level of potassium in the body is strictly controlled by our metabolism to maintain the normal range for biological functioning. Almost all of this is the stable isotope ^{39}K. However, 0.0117% of the potassium in the body is the long-lived isotope ^{40}K (half-life of 1.28×10^9 years). It emits both beta particles (89% of the time) and gamma rays (11%), with an average energy of about 1.5 MeV. It remains at a stable value of 1.71 nCi per kg in the human body.

^{14}C is another long-lived isotope that occurs in many foods. It is produced by the bombardment of nitrogen nuclei in the upper atmosphere by cosmic neutrons via (^{14}N + n $\rightarrow$ ^{14}C + p). ^{14}C is a very small fraction (~ 1.3 x 10^{-12}) of the stable ^{12}C, with a half-life of about 5,730 years. It remains in the bones of the body for a long time (where it provides the means of Carbon

Table 5.2. Radioactivity of some foods [UNSCEAR, 1982: Klement, 1982: IRSN,2012].

Food	^{40}K (pCi/g)	Food	^{14}C (pCi/g)
Nuts (Brazil)	5	Milk	0.5
Carrot	3	Poultry	1.8
Potatoes	3	Vegetables	0.5
Beer	0.4	Fish	0.5
Whisky	1.2	Grains	2.2

dating). ^{14}C emits only very low energy beta particles, of average energy around 49 keV, so it contributes less to the overall radiation dose to the body than does ^{40}K. Table 5.2 shows some approximate.representative natural activities in food.

Of course, we also receive radiation from other people's bodies—about 20 μSv per year from sleeping next to someone for 8 hours every night!

5.1.2 *The artificial radiation background*

The global average background radiation from man-made sources contributes about 20% of the total annual effective dose worldwide, though up to 50% of the total in developed countries, where the contribution of medical procedures has increased sharply over recent years. In North America in 1980, for example, medical exposures were about 0.6 mSv out of a total of 3.1 mSv; by 2006 they had reached 3.1 mSv out of a total of 6.2 mSv. [UNSCEAR, 2010].

Figure 5.2 indicates the dominance of diagnostic and nuclear medicine. The increasing contribution of medical procedures is somewhat offset by improved safety procedures and more efficient equipment. However it remains difficult to make general statements about the dose from specific medical and dental X-ray procedures, since the age

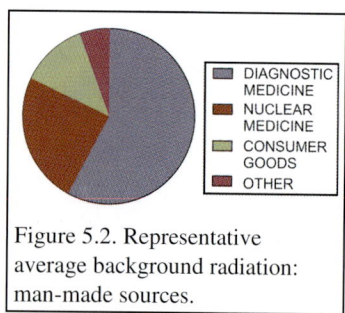

Figure 5.2. Representative average background radiation: man-made sources.

and quality of the equipment used is widely variable. In addition, the radiation burden varies widely from person to person and from procedure to procedure. A search of the literature often yields estimates that differ by as much as factors of ten.

A typical chest X-ray exposes the patient to an effective dose of about 50μSv to 100 μSv; multiply these values by up to twenty for a Barium enema exposure or a CT scan. The effective dose of a (bitewing) dental X-ray can be as little as 5 μSv to as high as ten times that amount. A panoramic dental X-ray is equivalent to about 9 individual bitewing exposures.

Consumer goods such as colour television sets, video displays of various sorts, smoke detectors (that use Americium, a radioactive isotope) account for most of the rest. Unsurprisingly, this source of background reaches its greatest value in developed nations.

Other odd sources that may expose you to radiation include reading a book for 3 hours per day (about 10 μSv per year due to small amounts of radioactive materials in the wood used to make the paper), flying in an airplane (about 50 μSv for a return trans-continental flight—the level of radiation exposure doubles with each 6,000 foot increase in altitude and adds about 10 μSv of radiation for each 1,000 miles you fly), smoking a pack of cigarettes a day (this is a serious one!) which produces about 5 mSv of radiation per year from the high concentration of ^{210}Po, a product of the radium decay chain, which attaches itself to the tobacco leaves.

5.2 Radiation—risk or benefit?

Exposure to high doses of ionizing radiation above the background is undoubtedly bad for your health. This section will give you some understanding of the complex and often controversial relationship between exposure to radiation and human health.

It is convenient to discuss separately 'low' dosage and 'high' dosage, though the line between the two is fuzzy, depending on a variety of factors including the radiosensitivity of the individual and of different cells. For the purposes of discussion, we will discuss dosages above and below 1 Sv; however, UNSCEAR(2010) considers high dosage as anything above about 200 mSv. Growing young people are more at risk, since the more rapidly a cell is dividing, the greater is its radiosensitivity—which is, of course the reason that radiation preferentially attacks cancer cells. Dose rate is also important; obviously a dose received in a fraction of a second (e.g from the atomic bomb dropped on Hiroshima) will have a very different effect from the same dose received over a year (by, for example Ontario miners). The discussion below over-simplifies a complex subject. Unless otherwise indicated, the doses refer to whole body dose.

5.2.1 *Symptoms of exposure to high radiation dosage*

A high dosage is defined as above about 1 Sv. People exposed to a single dose of 1 to 2 Sv suffer nausea, loss of appetite, and possible redness of the skin within a short time (several hours). Recovery takes several weeks. For whole body doses of from 2 to 4 Sv, the onset of radiation sickness is more rapid and more severe, and recovery takes many months if it occurs at all. There may be bleeding and infection, vomiting, hair loss, and sterility. Above about 4 or 5 Sv, death is almost certain.

The risk from high dosage is called 'deterministic', since a direct link can be made between the exposure and the symptom. Of course, if the individual survives, risks of the sort discussed in the next section will also be incurred. Figure 5.3 shows the non-linear relationship between deterministic effects and dose in this region.

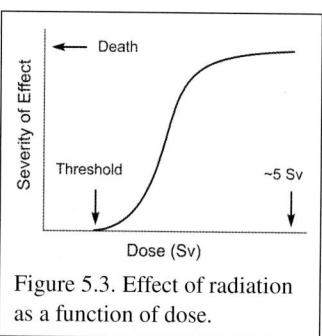

Figure 5.3. Effect of radiation as a function of dose.

The results of the atomic bomb attacks on Hiroshima and Nagasaki at the end of the Second World War, which exposed thousands of people to high energy gammas and neutrons, might be thought to provide useful data; however, most people in the high dosage areas were killed by the mechanical blast, not by the radiation. Most survivors were located beyond 1 km from the central explosion where the radiation dose had dropped to around 1 Sv. Data from these survivors provide information on the risks from low doses (see §5.2.2).

The nuclear disaster in 1986 at Chernobyl, the Ukranian nuclear reactor, sent a huge radioactive cloud around the world, and rendered the immediate area unfit for human habitation for the foreseeable future.

A comprehensive report was issued in 2005 by the Chernobyl Forum, a consortium of eight specialized international organizations [Chernobyl Forum, 2005]. The radioactive fallout from Chernobyl affected 5 million people, mainly in Ukraine, Belarus, and Russia; with the exception of the 1000 or so people who received immediate high doses of radiation, most of the others received dosages only slightly in excess of the natural background. Within the population at high risk, who received doses between 2 and 20 Gy, acute radiation syndrome was diagnosed in 134 emergency workers. Approximately 50 died from this exposure. No good data exists for the doses received by these workers, since many of the dosage meters that survived the explosion were found to be inaccurate. Table 5.3 shows an estimate of the fate of 115 of those most severely exposed (mainly firefighters).

Table 5.3. Casualties from the Chernobyl disaster [McNeill & McNeill, 2000].

No. People Exposed.	Received Dose (Sv)	Number of Dead
31	1 – 2	0
43	2 – 4	1
21	4 – 6	7
20	> 6	20

In 2011, a tsunami destroyed reactors at the nuclear power plant in Fukushima, Japan, leading to the worst nuclear disaster since Chernobyl. However, since the exposure of local populations to radiation was less from Fukushima, the risks are expected to be lower. There have been no cases of radiation sickness or death from high radiation exposures.

As discussed in Chapter 4, therapeutic radiation doses are often of the order of several Sv. These are given under carefully controlled conditions, almost always concentrated to small volumes of the body, with each exposure lasting only a short time. After each exposure, time is given for the cells to recover before the next one is administered.

5.2.2 *Risks of exposure to low radiation dosage*

Below whole body doses of about 1 Sv, typical of diagnostic exposures, most effects are not immediately obvious. A single dose of above 250 mSv produces a drop in blood count, which is, however, quickly reversible. The longer term risks, being mainly **carcinogenesis** (the production of cancers) or **mutagenesis** (the mutation of cells of the body), are more difficult to calculate. These risks are called '**stochastic**' (meaning 'random'); though it is believed that certain doses cause cancer in a certain proportion of the exposed population, no prediction can be made about which individuals will contract the disease.

The calculation of the stochastic risks associated with low annual doses (i.e. those similar to or less than the naturally occurring background) is difficult and controversial. The difficulty is that relatively few people receive annual doses greater than the normal background radiation, and deleterious health effects are observed only at very much greater doses.

Atomic bomb survivors who received doses in the 100 to 200 mSv range and above had elevated risks, with greater incidences of leukemia and solid cancers than the general population [UNSCEAR, 2010].

There is now considerable data from the Chernobyl nuclear accident on the exposure of people to increases in the background radiation. Local populations have been exposed to an annual exposure over the normal background of anywhere from 1 mSv to around 50 mSv. However, any increase in cancer rates are impossible to extricate from the normal occurrence of cancer from radiation and all other causes.

One of the greatest risks accrue to people who were exposed when they were children to radioactive ^{131}I, mainly from drinking the milk of cows that grazed on contaminated pastures; the risk of thyroid cancer rises with the dose received. By 2002, more than 4000 cases of thyroid cancer have been diagnosed from people exposed to radiation from the Chernobyl accident. Again, however, it is impossible to make a direct link between these cases and specific values of excess radiation dosage.

The Chernobyl Forum report [Chernobyl Forum, 2005] projects that "... *among the most exposed populations ... total cancer mortality might increase by up to a few per cent owing to Chernobyl related radiation exposure.*" Up to several thousand fatal cancers might result, but the study concedes that such an increase would be very hard to detect over and above the 100,000 cancer deaths expected from all other causes.

In the case of Fukushima, the difficulties of extrapolation are even more difficult. According to a comprehensive report issued by the World Health Organization [WHO, 2013], local populations received an excess of 12 to 25 mSv in the first year after since the disaster. Ten Hoeve and Jacobson

(2012), give a best estimate of 180 cases of cancer directly caused; however the great uncertainties in these types of calculation lead to a range of uncertainty from 24 to 2,500! Another study [Tsubokura, 2012] finds detectable levels of radioactive ^{137}Cs in people living within a few kilometres of Fukushima, but the total dose received, less than 1 mSv, is not considered to be a health risk. The WHO report concludes that, although there are increased risks over the baseline lifetime rates for specific cancers in specific segments of the population, *"... for the general population inside and outside of Japan, the predicted risks are low and no observable increases in cancer rates above baseline rates are anticipated."*

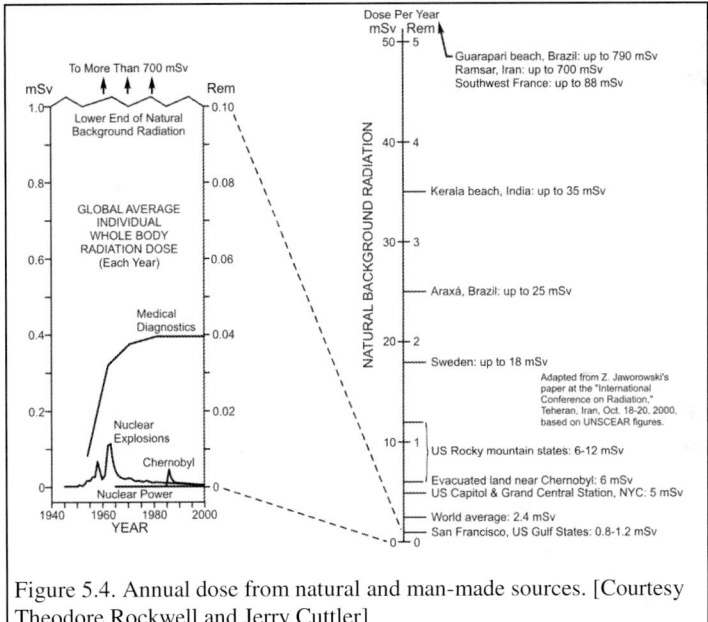

Figure 5.4. Annual dose from natural and man-made sources. [Courtesy Theodore Rockwell and Jerry Cuttler]

Although somewhat outdated, Figure 5.4 provides a clear and dramatic summary of the situation. The increased environmental radiation caused by nuclear power stations, the nuclear bomb tests in the 1960's, the nuclear power plant accidents at Three Mile Island in 1979 in the US, or Chernobyl in 1986, are seen to be small and temporary contributors to the overall average background. This does not, of course, deny the fact that ongoing effects on people in the close vicinity of the tests or the accidents are, in many cases, severe.

The huge variations in background. levels of radiation have been used to study the effects on human biology of low annual rates. No effect has been reported. Indeed one study shows that there is 15% *less* cancer in US states where the background radiation is *higher* than the US average [Jagger 1998]. Yet another high-statistics study was carried out in 1980 on two groups of 70,000 people each in China's Guangdong province [HBRRG, 1980]. One group had an annual radiation dose of about 1 mSv, while the other had about three times that amount, due to the different terrestrial environment. No short-term or long-term health differences were found between the two groups. Indeed, there has been no statistically significant effect demonstrated at doses below about 100 mSv at any dose rate.

The conclusion is clear: any attempt to correlate low dose rates of less than about several hundred milli-sieverts with health effects are doomed to failure.

Nonetheless, regulatory bodies that have the responsibility for setting safe limits on radiation exposure need a model to guide their recommendations. A popular model that extrapolates from high doses to very small ones is the Linear, No Threshold (LNT) hypothesis, which assumes that the risk of cancer increases linearly with exposure, starting from zero dose, and that there is no threshold below which radiation is safe.

This hypothesis suggests that the approximate overall risk coefficient for cancer and heritable effects is about 5×10^{-5} times the number of mSv received, or about 5% per Sv. This means that, for instance, a dose of 50 mSv is estimated to increase the probability of contracting fatal cancer by $5 \times 10^{-5} \times 50 = 0.0025$, or 0.25% over and above the "normal" risk of about 25%. An increase of 0.1 mSv in the dose you receive increases your

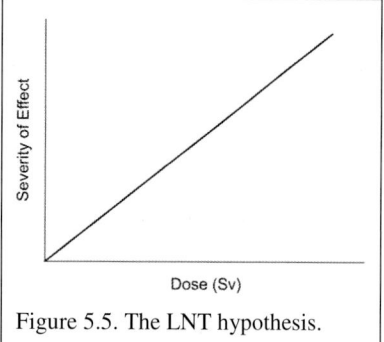

Figure 5.5. The LNT hypothesis.

chance of dying by 5 parts in a million, $5 \times 10^{-5} \times 0.1$, but so do smoking less than 2 cigarettes, spending 2 days in New York city, or travelling 40 miles in a car (the last two risks have many other contributing factors!).

Using the LNT hypothesis, the U.S. National Council on Radiation Protection and Measurements [NCRP, 1988], estimates that between 1 in 10 to 1 in 7 of all lung cancer deaths in the US (i.e. about 15,000 to 22,000 per year in the US) are caused by exposure to naturally occurring radon in the home which compares favourably with the the figure of approximately 21,000 from epidemiological studies. In comparison, around 50,000 Americans annually are killed in traffic accidents.

In most countries, regulatory bodies set limits on the amount of radiation allowable in the workplace; usually around 50 mSv in one year, and 100 mSv over five years. Additional regulations apply to pregnant women, since it is thought that potential damage is high for fetal cells. Particularly at risk are workers in medical or dental institutions where X-ray equipment is in use, and in the nuclear industry.

However, many experts have pointed out that the extrapolation from the known deleterious effects at high dosage, which are approximately linear, to very low doses is not justifiable. Many scientists claim that the LNT theory is extremely conservative, if not simply wrong. Jaworowski [1999] has pointed out that *"the fear of small doses ... is about as justified as the fear that a ... temperature of 20 °C may be hazardous because, at 200 °C, one can easily get third degree burns"*. Similar concerns have been expressed by Thomas, [1994]. Dr. Gunnar Walinder, the renowned Swedish radiobiologist (an associate of Rolf Sievert) even has a monograph entitled *Has Radiation Protection Become a Health Hazard?* [Walinder, 2000].

In summary, the weight of evidence seems to support the statement made by the Health Physics Society (2010) which recommends : *"... against quantitative estimation of health risks below an individual dose of 50 mSv in one year or a lifetime dose of 100 mSv in addition to background radiation. Risk estimation in this dose range should be strictly qualitative ...*" The report continues*: "There is substantial and convincing scientific evidence for health risks following high dose exposures. However, below 50–100 mSv (which includes occupational and environmental exposures), risks of health effects are either too small to be observed or are nonexistent.*"

There is even some suggestion that exposure to radiation may be hormetic (meaning that low doses have a beneficial effect, while higher doses may be harmful). While this concept has not been endorsed by any of the regulatory radiation agencies, it has not been disproved, and does have its adherents. Several old gold and uranium mines in Montana even boast of the health advantages of low levels of radiation, and, for a fee, you can sit in the mines to breathe in the radon gas and drink the radioactive water! Boreham [2002] of McMaster University has shown that developing mouse fetuses are protected by prior exposure to low doses of radiation against the effects of *in utero* irradiation that interfere with normal embryonic development.

Exercises Chapter 5

1. Another unit used in some radiation risk studies is the Background Equivalent Radiation Time. It is defined as the time that would give an adult the same Effective Dose as they would receive from the background radiation. Calculate the BERT in days for a chest X-ray.

2. Another unit is the Banana Equivalent Dose (BED) which is the dose received by a person who has just eaten a banana! The BED is supposed to provide a simple standard against which other exposures to radiation may be estimated. As can be seen from Table 5.1, potassium is the main contributor to radiation from our bodies (the contribution from the next radioisotope, ^{14}C, is much smaller and can be neglected). The level of potassium in the body is strictly controlled by our metabolism to maintain the normal range for biological functioning, so any extra ingested potassium is excreted within hours. 0.0117% of the potassium in the body is the long-lived isotope ^{40}K (half-life of 1.28×10^9 years). It decays with an average energy of about 1.5 MeV. A banana contains about 0.5g of potassium. Suppose

an "average" 70kg adult eats a banana, and his or her body returns to homeostasis by excreting the extra potassium within 12 hours. What is the value in Sv of the BED calculated on this data?

3. Nuclear tests in the atmosphere released into the atmosphere a quantity of radioactive ^{137}Cs, a product of the uranium fission chain; thanks to the high solubility of Caesium's chemical compounds ^{137}Cs appeared in the world's drinking water. It has a half-life of about 30 years and emits a gamma ray of 0.66 Mev. Before nuclear weapons tests were outlawed, people had absorbed radioactive ^{137}Cs resulting in an activity of about 40 Bq. A) What mass of ^{137}Cs does this correspond to? B) What is the lifetime effective dose received by an average 70 kg man from this source? (State the assumptions you make).

4. Using the LNT hypothesis, calculate the approximate probability of eventual death from radiation, over and above the normal risk of about 25%, for A) a dose of 50 rem? B) a chest X-ray? C) a person who smokes a pack of cigarettes every day for 40 years? (Hint: start with the estimates given in §5.2.2.)

5. The maximum permissible dose for workers using X-rays is 5 rem (50 mSv) per year. (Assume a 40 hour week, with a three-week holiday period per year.) What is the safe working distance from a ^{60}Co source that produces 0.3 rad (3 mGy) per hour at a distance of 1m.

6. In April 1986, there was an explosion of a reactor at the Chernobyl Nuclear Power Plant in the Ukraine (then part of the USSR). A large cloud of radioactive elements entered the atmosphere, affecting 5 million people in the USSR, Europe, and even North America. It is estimated that over 10,000 people died of cancer caused by the radiation. Nearby Prypyat, once inhabited by 47,000 people, is now a ghost town. The radioactivity of milk in many regions rose to 2000 Bq.L^{-1} due to the Iodine-131 in the contaminated grass eaten by cattle. With a half-life of about 8 days, ^{131}I is particularly dangerous, as Iodine concentrates in the thyroid gland ; indeed children exposed to this source show an increase in thyroid cancer. Of course, normal milk, which contains potassium, is mildly radioactive due to the potassium it contains (potassium is, of course, one of milk's health benefits). 1 litre of milk contains about 2.00 g of Potassium, of which

0.0117% is the radioisotope ^{40}K. A) How long after the accident at Chernobyl would the activity of the ^{131}I in the milk have fallen to the 'natural' level of the ^{40}K? B) A young girl drank half a litre glass of contaminated milk. What was the dose to her thyroid gland from this one glass? What was the effective dose she suffered? Assume the mass of her thyroid gland is 15g and that all the iodine in the milk concentrated in her thyroid. For the purposes of this calculation you may assume that the average energy of the several γ and β decay products of ^{131}I averages to about 600 keV. C) If she drank a glass of this milk every day, what was the average dose rate delivered to her thyroid after a week? You may ignore biological excretion, but allow for nuclear decay. What was the average effective dose rate after a week? D) Comment on the health implications.

7. On the 1st of November 2006, an ex-KGB agent called Alexander Litvinenko, who had sought political asylum in the UK, was taken to hospital. 22 days later he was dead. Before he died he accused Russian president Vladimir Putin of having engineered his death. Exhaustive tests on Litvinenko established that he had been poisoned by a radioactive isotope, Polonium-210. The Guardian newspaper states that the amount of Polonium-210 found in the Russian's body would have cost as much as £20m (~ $50 million) to acquire.

Polonium-210 was discovered by Marie and Pierre Curie in 1898, who named it after Marie's native Poland. It occurs naturally in uranium ores, being one of the decay products of uranium, and is made artificially at nuclear reactors. It also appears in tobacco plants which have been fertilized by calcium phosphate; one cigarette contains around 0.04 pCi and the US Surgeon General has claimed that this radioactivity, rather than the tar, accounts for 90% of all smoking related cancers. ^{210}Po is five times more deadly when inhaled than ingested. ^{210}Po decays with a half-life of 138 days by emitting 5.41 MeV alpha particles, which are easily stopped by skin, or by a few centimeters of air. However, when ingested, the soluble Polonium circulates to every cell and tissue in the body, where the alpha particles deposit all of their energy. The half-life for biological excretion is about 50 days. Comment on the accuracy of the numbers in the following quotes of the time:

a. The average American receives about ¼ Gy in a lifetime from cosmic rays, the soil, and medical exposure (*USA Today*)

b. ^{210}Po is about five thousand times as radioactive per gram as Radium (*Toronto Globe & Mail*)

c. 1mCi of Polonium would weigh only 0.2 millionths of a gram (*Toronto Globe and Mail*)

d. 3 mCi of ^{210}Po would be sufficient to kill (*Health Physics Society*)

e. A fatal 4-Sv dose can be caused by ingesting 8MBq (200 microcurie), about 50 nanograms (ng) of ^{210}Po (*Wikipeidia*).

Chapter 6

Magnetic Resonance Imaging

6.1 Introduction: NMR and MRI

Magnetic resonance imaging, or MRI, is one of the fastest growing and most powerful technologies used in medical diagnosis and research. The physics of Nuclear Magnetic Resonance (NMR) was developed as early as the 1940's by Felix Bloch and Edward Purcell at Harvard University, for which they won the 1952 Nobel Prize. However, it took the computer revolution to allow the procedure to become one that is now used routinely for diagnosis and medical research. The applications of MRI grow by the year; one of the most interesting recent ones, called functional MRI (fMRI), allows researchers to study which different parts of the brain are activated as a volunteer subject undertakes various tasks.

6.2 The physics of nuclear magnetic resonance

6.2.1 *The spinning top in a gravitational field*

We will start by using an example of a macroscopic object that may help you to understand the more complex quantum mechanics that provides the basic mechanism for NMR and MRI. When a top is set spinning so that its spin axis makes an angle with the vertical, it will rotate around the vertical, in such a way that its topmost point will trace out a circle; the axis of the top will maintain

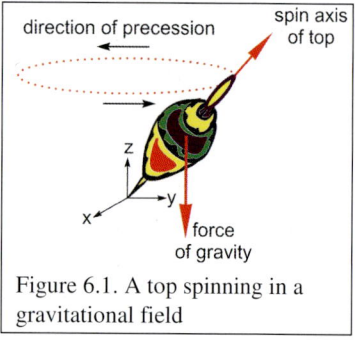

Figure 6.1. A top spinning in a gravitational field

the same angle to the vertical (until friction begins to slow it down). This is what is meant by '**precession**'; the top is said to precess around the vertical.

It turns out that electrons, protons, and many nuclei also have a quantity which looks a lot like spin, although, as you might expect by now, this type of spin is quantum-mechanical in nature, and behaves in some startlingly different ways.

6.2.2 *Quantum-mechanical spin*

We have seen in Chapter 1 how the existence of different electronic energy levels in atoms leads to line spectra. Careful experiments showed, surprisingly, that some of these lines are split, with differences in wavelength that are much smaller than could be explained by any known differences in the atomic energy levels.

In 1925, George Uhlenbeck, and Samuel Goudsmit explained this 'fine splitting', by hypothesizing a new quantum number for the electron that they called 'spin'. Spin is an entirely quantum mechanical phenomenon; although it has units that are the same as classical angular momentum— contributing small positive or negative amounts to the atomic energy levels—it behaves in distinctly non-classical ways.

The spin of an electron is also called its intrinsic angular momentum, which distinguishes it from the orbital angular momentum it has as it 'orbits' around the nucleus in an atom. Its value is $\sqrt{s(s + 1)}\hbar$ where $s = \frac{1}{2}$, with $\hbar = h/2\pi$ (h is Planck's constant: §1.6). Often we will simply say that 'the spin of the electron is $\frac{1}{2}$'. The proton, neutron, and electron have spins (or intrinsic angular momenta) of the same magnitude; all these particles are called 'spin $\frac{1}{2}$ particles'. The proton spin produces a corresponding magnetic moment, denoted by μ (not to be confused with the linear attenuation coefficient of §2.5!) that points in the same direction as the spin; protons behave in some ways like small magnets.

The magnetic moment of a proton is directly proportional to its spin and is given by $\mu = \gamma(\hbar/2)$, where γ is called the **gyromagnetic ratio**. Its value has been measured to be:

$$\gamma = \mu/(\hbar/2) = 2.68 \times 10^8 \text{ s}^{-1}\text{T}^{-1} \tag{6.1}$$

6.2.3 Magnetization of the sample

In an un-magnetized sample of a material containing 'mobile' protons (i.e. those not too tightly tied to atoms or molecules), the spins of the protons point in random directions, as, therefore, will the magnetic moments; the sample will have no net magnetic moment. However, when a constant magnetic field, B_0, is applied (let us suppose in the +z-direction, so that $B_z = B_0$), we might expect all of the magnetic moments to line up in exactly the same direction, like small magnets. However, this is not exactly what happens.

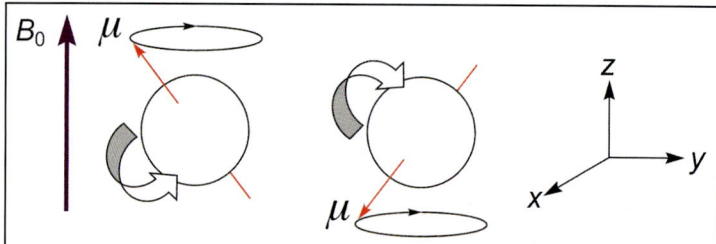

Figure 6.2. Spinning protons in a constant magnetic field in the +z direction. The direction of the magnetic moments, which is that of the proton spins, is indicated by the red arrows labeled μ. [Courtesy Keevil, 2001: adapted].

The quantum mechanical nature of the proton allows the value of the z-component of the spin in the direction of the magnetic field, s_z, to have two, and only two, values: $+\hbar/2$ (in the direction of the field) and $-\hbar/2$ in the opposite direction as shown. Thus the z component of μ can also have only two values.

In addition, the protons precess around the +z axis (in a similar fashion to the spinning top shown in Figure 6,1). The angular frequency of this precession can be calculated using classical physics: it is called the

Larmor frequency (named after Joseph Larmor), given by $\nu_L = \omega_L / 2\pi$, where $\omega_L = \gamma B_0$, with γ the gyromagnetic ratio. Substituting values, the Larmor frequency is, with B_0 in T and ν_L in MHZ:

$$\nu_L \cong 42 B_0 \qquad (6.2)$$

Of the two possible orientations for the direction of the magnetic moments of the precessing protons the one with the z-component of spin opposite to the direction of the magnetic field has the greater energy (since energy from the field has to be supplied to hold it in place). We might therefore expect that all of the protons orient themselves in the other, lower energy state in the +z-direction, as small magnets would. However this does not happen, because of their random thermal motion.

Our old friend Boltzmann calculated this effect in the distribution that bears his name: the number of particles occupying a state of energy E is proportional to $exp(-E/kT)$, where k is Boltzmann's constant, and T is the temperature in Kelvin. Thus there is an excess of protons aligned with the magnetic field, which have less energy than those aligned against it; this excess is small, of the order of a few parts per million (§6.2.4). However, the huge number of protons in even a small amount of material means that the contribution of the unpaired protons is enough to produce a measurable net macroscopic magnetic moment, M, in the field direction (i.e. in the presence of B_0 alone, $M_z = M$). See Figure 6.3.

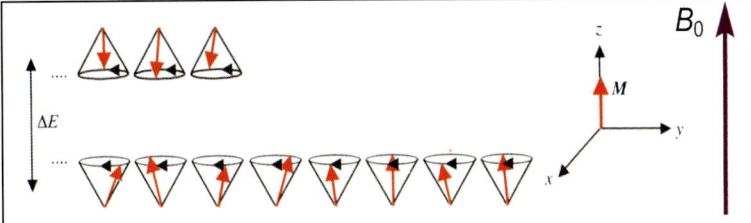

Figure 6.3. Schematic of the spinning protons aligned with (bottom row) and against (top row) the magnetic field B_0 applied in the +z direction. [Courtesy Keevil, 2001: adapted].

Since the protons do not, in general, precess in phase, the x- and y-components of the magnetic moments are randomly directed, and the sum of these components over many protons averages to zero. Defining

M_{xy} as the magnitude of the magnetic moment in the x-y plane, this situation can be represented by $M_{xy} = 0$.

6.2.4 *Energy considerations*

Let the energy of the n_+ protons pointing "up" (in the +z direction) be E_+, and that of the n_- protons pointing "down" be E_-, so that the energy difference is $\Delta E = E_- - E_+$. This energy difference can be expressed in terms of the proton magnetic moment and the strength of the magnetic field (the classical calculation is shown in Appendix A6.1); the result is

$$\Delta E = E_- - E_+ = 2\mu B_0 \tag{6.3}$$

For a typical magnetic field, ΔE is small ($\approx 10^{-7}$ eV for a 1Tesla field). Using the Boltzmann expression given in §6.2.3 above, the ratio of the numbers of protons in the lower energy state to those in the higher energy state at a temperature T is given by:

$$n_+/n_- = exp(\Delta E/kT) \tag{6.4}$$

The difference between n_+ and n_- (i.e. the number of unpaired protons) is thus several parts per million.

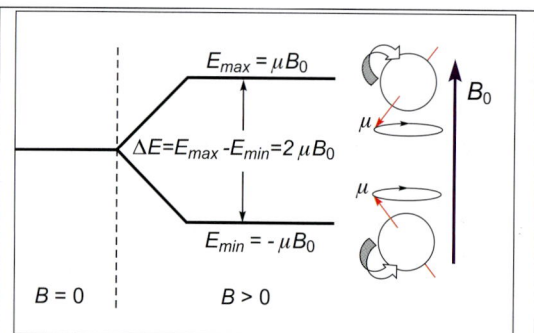

Figure 6.4. The energies of the two proton states in the presence of a field, B_0 [Courtesy Keevil, 2001: adapted].

The difference in energy between the two states provides a possible method of measurement. If energy of exactly ΔE is directed into the material, it can be absorbed by protons in the lower energy state causing them to jump into the higher state. The source of this energy is conveniently provided by an oscillating magnetic field, B_1, called the **rf field** (the Larmor frequency and other frequencies used in MRI lie in the radio-frequency—rf, or FM—domain; see Figure 1.8).

As indicated in §1.6, a magnetic field, oscillating at a frequency of v, can be treated as a source of photons of energy $E_\gamma = hv$. When the frequency of this field is tuned so that E_γ exactly equals ΔE, a **resonant condition** is said to have been obtained. At resonance, the absorption of the field's energy shows a sharp maximum. The frequency at which this happens, for reasons that will become clear in the following sections, is just equal to the Larmor frequency, v_L, the natural frequency of precession of the protons in the main constant magnetic field B_0.

The following equation summarizes different expressions for the energy difference between the two states of the protons in a magnetic field:

$$\Delta E = hv_L = (h/2\pi)\omega_L = \hbar\gamma B_0 = 2\mu B_0 \tag{6.5}$$

6.2.5 *Measuring the Larmor frequency*

In addition to allowing a measurement of the Larmor frequency, the application of the small oscillating field, B_1, yields a wealth of useful information. This section gives a simplified description of the dynamics.

Consider a small sample of material placed in a strong, homogeneous, and constant magnetic field $B_0 = B_z$. As described above, the 'mobile' protons will align with and against B_0. The small number of these that are unpaired will produce a constant magnetic moment also pointing in the +z direction, $M = M_z$. Since the proton magnetic moment is known, a measurement of the magnitude of this macroscopic magnetic moment, M, would provide information on the number of unpaired mobile protons in the sample, i.e. $n_+ - n_- = M/\mu$. However, the small field from M is completely masked by the large applied field, B_0, which points in the same direction. In order to measure M, it is necessary to rotate it away from the direction of B_0 (i.e. away from the z-axis) so that its effects are observable. If the frequency of B_1 is tuned to equal the Larmor frequency, the required rotation can be achieved.

The Larmor frequency is the natural frequency of precession of the spinning protons. When the external B_1 field is applied at exactly this frequency, the protons begin to precess in phase with the field and with each other. (This resonant condition is analogous to that obtained when pushing periodically on a child's swing; when the period of the push equals the natural period of oscillation of the swing, the amplitude of the swing increases easily.) The net effect is that the macroscopic magnetic moment, being the sum of the magnetic moments of the spinning protons, acquires x- and y- components, and spirals down around the z-axis under the forcing effect of this applied field. It can be shown that the angle, θ, between M and the z-axis increases from zero with time, τ, according to the equation $\theta = \gamma B_1 \tau$. For purposes of discussion, suppose that B_1 is applied long enough for M to reach the x-y plane, where $M_z = 0$. The applied field is then switched off. In this case, θ equals $\pi/2$, and the B_1 pulse is called a 90° rf pulse. In practice, diagnostic MRI uses a wide variety of pulses, each with different effect. The macroscopic magnetic moment is now rotating at the Larmor frequency in the x-y plane.

As soon as the oscillating field B_1 is switched off, the magnetic moment begins to spiral back to its equilibrium position pointing along the +z-axis. According to Faraday's law, the oscillating magnetic field from this rotating magnetic moment produces an electric field that generates an emf in a coil of wire placed in its vicinity. This emf has a frequency equal to the Larmor frequency, and its strength allows a determination of the magnitude of M. This behaviour is called the **relaxation** of the moment. As discussed in more detail below, this relaxation reduces the x-y components to zero, and re-establishes the value of M_z to its maximum of M. As a bonus, a measurement of the time it takes for this relaxation—the relaxation time— yields information about the nature of the material being studied (see §6.3.2)

The excitation of the spins in the sample by a 90° rf pulse of B_1 and the subsequent relaxation are visualized in the figures 6.5 to 6.7.

This is the starting position. The main homogeneous magnetic field B_0 has been switched on. An excess of protons with precession vectors pointing in the direction of B_0 creates the macroscopic magnetic moment M, that also points along the +z axis.

$$M_z = M \text{ and } M_{xy} = 0$$

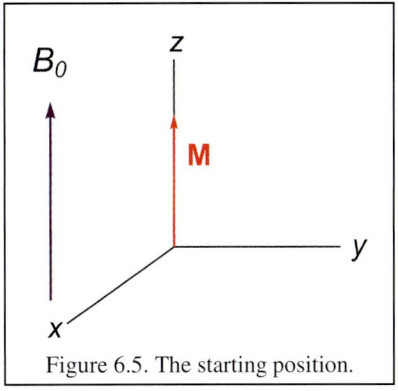

Figure 6.5. The starting position.

The oscillating field, B_1, is now switched on. The magnetic moment vector M spirals down to the x-y plane.

$$M_z \to 0 \text{ and } M_{xy} \to M$$

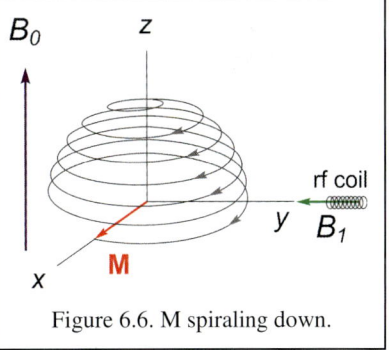

Figure 6.6. M spiraling down.

The B_1 field is now switched off. As the magnetic moment vector returns to its original position—relaxes—the signal from the magnetic moment as it sweeps around in the x-y plane is detected by the pickup coil.

$$M_z \to M \text{ and } M_{xy} \to 0$$

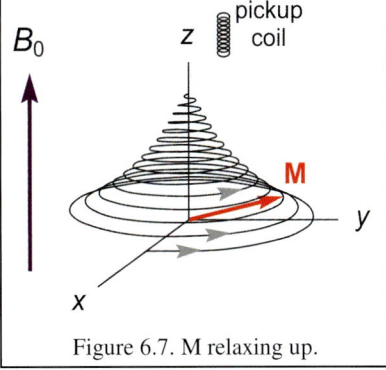

Figure 6.7. M relaxing up.

The rf signal (called the Free Induction Decay, FID) from the pickup coil is shown in figure 6.8; it oscillates at the Larmor frequency, and decays with time as the x-y component of M (M_{xy}) relaxes to zero.

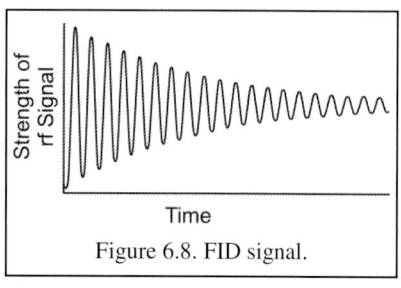

Figure 6.8. FID signal.

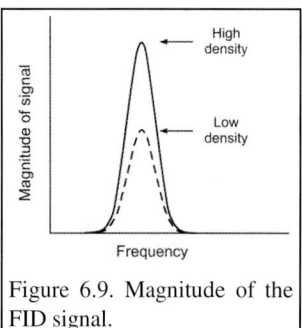

Figure 6.9. Magnitude of the FID signal.

The decay time of the rf signal from the pickup coil, obtained from the plot in figure 6.8, depends on the properties of the material being studied, and allows discrimination between different tissues (§6.3.2). The *magnitude* of the signal, shown in Figure 6.9 is proportional to the number of mobile protons in the sample.

In summary, the magnitude of the magnetic moment, M, is proportional to the density of protons in the sample of material, which is, in turn, related to the composition of the material. By disturbing M from its equilibrium value, and measuring its magnitude and the speed with which it returns to equilibrium, information about the material can be obtained. ***This is the basis of MRI.***

6.3 Magnetic resonance Imaging (MRI)

The physics of NMR, described in §6.2 above, is the basis for MRI, one of the fastest growing methods of medical diagnosis (the medical procedure dropped the word 'nuclear' to avoid making patients too anxious!).

NMR signals can be obtained from any nucleus that has a non-zero spin. Of the most abundant elements in the human body (H,O,C,N), only H and N satisfy this requirement; H is about 60 times more abundant than N. Since about 55% to 60% of the human body is water, it is the nucleus of the H atom in water—the proton—that gives the strongest signal.

MRI uses the principles of NMR to detect the density of protons in the water in biological tissue. The brilliant suggestion of Paul Lauterbur (a chemist by training), and the work of Peter Mansfield (a physicist), whose analysis took advantage of the powerful computers and analysis software used in X-ray tomography, allowed the development of MRI as a diagnostic tool that provided much better discrimination between soft tissues than X-rays, with no radiation risk. Lauterbur and Mansfield won the 2003 Nobel Prize for their work on MRI, although the first MRI image was taken by Raymond Damadian (a physician), who was also the first to realize that different tissues give different MRI signals. Figure 6.10 shows Damadian sitting in his first successful MRI machine. The first clinically useful information obtained from a whole-body MRI machine was obtained by an international team of researchers working at the University of Aberdeen, Scotland, in August 1980.

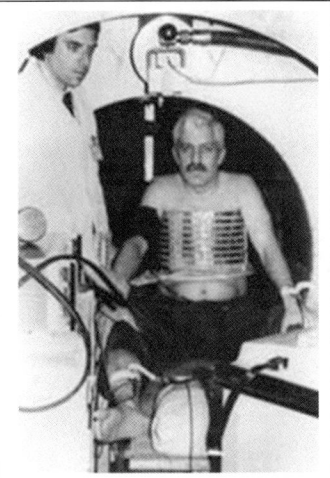

Figure 6.10. Dr. Raymond V. Damadian sitting in *Indomitable*, the world's first MRI scanner that he and his colleagues built. [Courtesy Dr Damadian and Fonar corporation].

6.3.1 *Lauterbur's brainwave—the gradient field*

An NMR signal from a human body placed in a homogeneous magnetic field, B_0, will yield a signal from the pickup coil whose magnitude depends on the total number of protons in the body, as described in §6.2.5 above. However, this signal will come from all parts of the body at once; there is no spatial discrimination. In order to identify the spatial position from which the rf signal is emitted, Lauterbur made the simple yet brilliant suggestion of adding a small linear gradient to B_0.

Suppose B_0 is aligned along the long axis of the body which we will call the z axis, as shown in Figure 6.11. Let a magnetic field gradient be imposed in the +z direction, directly proportional to the z coordinate; then the Larmor frequency of precession, which in turn is directly proportional to the magnetic field (Equation 6.2), is a simple function of the z position in the body. A measurement of this frequency allows a determination of the z coordinate of a slice of the body.

Let the gradient be G_z so that the total field in the z direction is

$$B_z = B_0 + zG_z \qquad (6.6)$$

Then the Larmor frequency, directly proportional to the field, is a linear function of the z-position;

$$\omega_L(z) = \gamma B_z = \gamma(B_0 + zG_z) \quad (6.7)$$

or, $z = \{\omega_L(z)/\gamma - B_0\}/G_z \quad (6.8)$

By measuring ω_L and knowing γ, B_0 and G_z, z can be determined. The magnitude of the signal at that frequency yields the number of protons in this slice at position z.

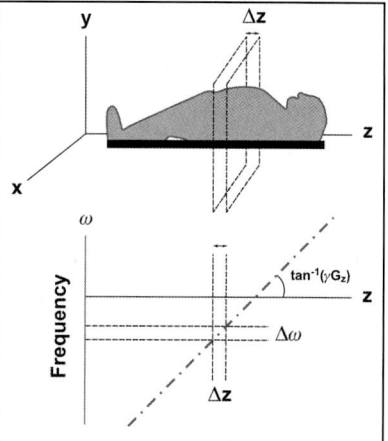

Figure 6.11. Schematic view of a patient in a magnetic field with a gradient in the z-direction.

The frequencies are measured in the manner described in §6.2.5; a short pulse of rf field, B_1 is applied at right angles to B_0; the magnetic moment relaxes, and its magnitude is measured by the emf it generates in the pickup coil. If this field has a bandwidth—a range of contained frequencies—of $\Delta\omega_L(z)$, protons whose Larmor frequencies lie in that range will be excited. Using equation (6.8) the width of the slice containing these protons, Δz, is given by:

$$\Delta z = \Delta\omega_L(z)/(\gamma G_z) \qquad (6.9)$$

The bandwidth of the applied rf field B_1 determines the width of the slice in which the protons are excited; the narrower the frequency range the narrower the slice to be examined, and the higher the spatial resolution.

Finally, the strength of the signal within $\Delta\omega_L(z)$, measured by the pickup coil, determines the number of protons in the slice of thickness Δz at position z. This procedure, of selecting the slice from which a signal will be received, is (not surprisingly!), called a slice selection.

Thus by sorting the overall signal from the pickup coils into different frequencies, the position from which that signal came can be identified, and the number of protons in different slices can be picked out. The idea is easily generalized. By applying field gradients at right angles to the first, in the x- and y- directions, the number of protons in each small volume in each slice can be determined. These small volumes are called 'voxels', by analogy to two dimensional 'pixels'. Computational methods, similar to those used in CT scans, are then used to correlate the measurements of proton density and relaxation time with the exact position in the body. In practice, gradients at many different angles are used to increase the resolution of the signals.

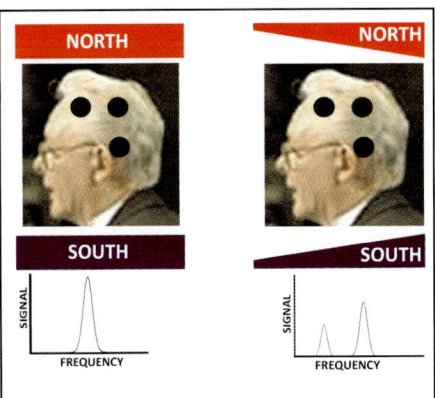

Figure 6.12. Schematic view showing the use of magnetic field gradients in the discrimination of position. [Courtesy Hornak, 2011: adapted].

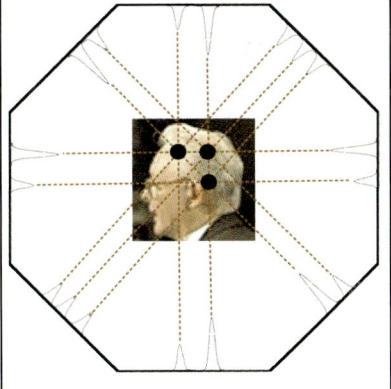

Figure 6.13. Location of voxels using a variety of magnetic field gradients at different angles. [Courtesy Hornak, 2011: adapted].

Figures 6.12 and 6.13 give a visual representation of this process (Dr Peter Mansfield is the 'patient'!). The black dots represent voxels under study. For the sake of simplicity, this representation assumes that each voxel has the same volume and contains the same number of protons.

In the diagram on the left of Figure 6.12, the rf signal from the pickup coil appears at one frequency—the Larmor frequency that corresponds to the magnitude of the homogeneous magnetic field. On the right, the magnetic field has a gradient at right angles to the main field; the signals from the two voxels on the right, located in a region of higher magnetic field, yield a higher frequency than the voxel on the left. The magnitude of the signal in each case is proportional to the number of protons, and, in this case, to the volume of the voxels.

In figure 6.13 the intersection of rf signals produced from magnetic field gradients at a variety of different orientations indicates schematically how the three voxels can be precisely located. In modern practice, this so-called 'back-projection' method of imaging is replaced by other, more sophisticated methods.

6.3.2 Relaxation times

The use of magnetic field gradients provides information on the density of protons in different small volumes of the body, and some discrimination between different tissues is achieved; for instance tumours have a higher proton density than most bodily tissue. However, since this density is similar for a wide variety of bodily tissues, the measurement of other parameters is used to provide even better discrimination between them. These parameters are the spin-lattice and spin-spin relaxation times.

When the applied oscillating field, B_1, is turned off, the proton spins take a finite time to return to their original configuration, lined up with or against the direction of the constant magnetic field B_0. This time is the relaxation time.

There are two independent relaxation times, shown schematically in figure 6.14. T_1 corresponds to the loss of energy between the proton spins and the atoms in the lattice; this loss causes M_z to return to its maximum value of $M_z = M$. This is called the **spin-lattice** relaxation time. T_2, the **spin-spin** relaxation time, corresponds to the exchange of spins between pairs of

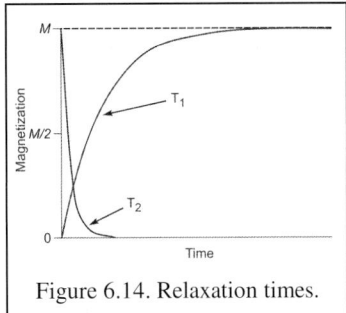

Figure 6.14. Relaxation times.

protons; this mechanism reduces M_{xy} to zero. The decay envelope of the FID signal shown in figure 6.8 corresponds approximately to T_2. Although these processes take place simultaneously, it is convenient to discuss them as if they were independent. T_1 is always greater than T_2. Both processes can be assumed to be exponential:

$$M_z = M\{1 - exp(-t/T_1)\} : M_{xy} = Mexp(-t/T_2) \qquad (6.9)$$

In 1970 Raymond Damadian showed that different tissues have different response times; in particular he was the first to realize that cancerous tumours have much greater relaxation times than healthy ones. Some representative values for healthy tissue are shown in Table 6.1. Using these parameters, differentiation between cancerous and healthy tissue can be as high as 180% (the figure for X-rays is only around 4%).

Table 6.1. Relaxation Times at 1.5 T for a healthy volunteer (error ± 10%) [Courtesy Keevil, 2001].

TISSUE	T_1 (ms)	T_2 (ms)
Grey Matter	1079	100
White Matter	684	79
Spinal Fluid	3959	914
Muscle	730	47
Liver	420	43
Fat	240	84

The measurement of T_1 and T_2 is complicated by several factors, such as unavoidable inhomogeneities in the magnetic fields. In practice, a complicated series of $90°$ and $180°$ rf pulses, separated by appropriate intervals, are used to provide the data necessary to measure T_1 and T_2.

6.4 Summary

The data-taking capability of an MRI system must be very great, since it receives rf input from the entire part of the body under study. The advent of MRI as a major diagnostic tool required the marriage of NMR to the development of computers and the sophisticated 3-D reconstruction software developed for CT scans (§2.5.2).

The MRI signal depends on several parameters: the strength of the signal provides information on the density of protons, the use of magnetic field gradients provides location, and the measurement of the relaxation times increases discrimination between different bodily tissues. The digital information is coded in a gray scale that depends on the location, the proton density, and the values of T_1 and T_2, so that a digital image can be constructed. Variation in these parameters allows good discrimination between those parts of the body that have high water content. Thus different soft tissues and tumours can be well differentiated (however, X-ray images are superior for examining bones). Contrast can be enhanced in different areas of the image, by choosing appropriate ranges of T_1 and/or T_2. Spatial resolution from MRI images depends on the time allowed for acquisition of the signal, but is typically in the millimeter range. Because of the time required to produce a good MRI image, X-ray techniques using real-time fluoroscopy are often more useful for observing bodily functions as they occur. However, new developments are announced regularly, and the field of fMRI is growing apace.

According to the Organization of Economic Cooperation and Development [OECD, 2011], there are now more than 20,000 MRI scanners worldwide; they cost anything from $1 million to more than double that, depending on the strength of the magnetic field. With 32.5 million people, Canada has around 6 units per million people, compared to 15 CT scanners. In comparison, per million people, Japan has 43 MRI (97 CT), the US has 26 MRI (41 CT), Denmark has 10 MRI (29 CT), and the UK has 6MRI (9 CT).

A6.1 *Potential energy of a magnet in a magnetic field*

Consider the work that we would have to do to move a classical magnet, of magnetic moment $\mu = mL$ from the position in which it is lined up with the magnetic field through 180°. The work done to move it from θ to $\theta+\Delta\theta$ is just the product of the torque, $F \times (L \sin \theta) = mB \times (L \sin \theta) = \mu B \sin \theta$, and the angular distance, $\Delta\theta$. The total work done can be found by integrating from 0° to 180°, to give the answer:

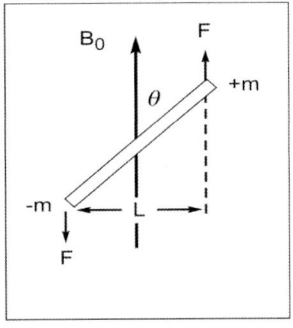

Figure A6.1. Classical magnet in a constant magnetic field.

$$W = \int_0^\pi (\mu B \sin \theta)\, d\theta = 2\mu B$$

Exercises Chapter 6

1. The gyromagnetic ratio of the proton is $2.67 \times 10^8\ \mathrm{s^{-1}T^{-1}}$. Calculate the magnetic moment of the proton.
2. Calculate the difference between the energies of the 'up' protons and the 'down' protons when placed in a magnetic field of 1.5 T. Express in eV. What is the relative number of 'up' protons expressed as a fraction of the total number of protons?
3. An MRI machine operates at 1.2 T main field, with a z gradient of $1.5\times10^{-2}\ \mathrm{T\ m^{-1}}$. A 90° rf field is applied. For how long must this field be applied and what must be the frequency spread to examine a slice of tissue of width 1 cm?
4. Many MRI machines operate at a magnetic field strength of 1.5 Tesla, though there are a few, mainly research machines that operate at 4.7 Tesla. Calculate the energies of the photons that will be absorbed by a ^{1}H nucleus in each field. How does this compare in energy to a 2×10^{19} Hz X-ray photon? What, approximately, is the ionization potential for a typical organic molecule? Which of the two photons will ionize the molecule?
5. A sample of hydrogen is placed in a 1.5 Tesla magnetic field. An rf field of $B_1 = 1.2 \times 10^{-4}$T is applied along the x-axis for 50

microseconds. A) What should its frequency be? B) What is the direction of the net magnetization vector after the B_1 field is turned off? C) If the B_1 field is doubled, how long must it be applied to produce the same effect?

6. A laboratory sample used for MRI research, containing three capsules of water (indicated by the three black dots in the diagram below), is placed in a magnetic field. The magnetic field points in the +z-direction, with a negative gradient in the +x direction such that $B_z(x) = B_0 - xG$ where G is a constant greater than zero, as shown in the diagram.

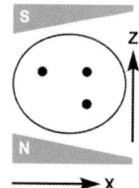

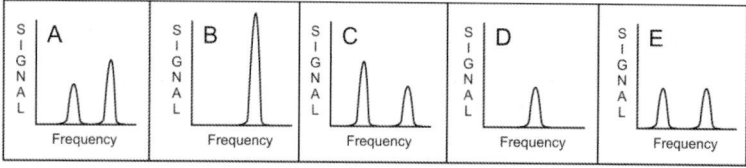

The graphs show the result of the measurement of the Larmor frequencies of the protons in each capsule. The values of frequency are plotted along the x-axis, and the strength of the signal is plotted on the y-axis. Which of the graphs most closely approximates the frequency spectrum you would expect to see?

7. In an MRI study of an obese patient, a measurement of the relaxation of the longitudinal magnetization, M_z (the z component of the macroscopic magnetization), for a small volume of the patient's body, yields the data shown in the graph. Which relaxation time does this graph represent, and what is the most likely constituent of the small volume of the patient's body?

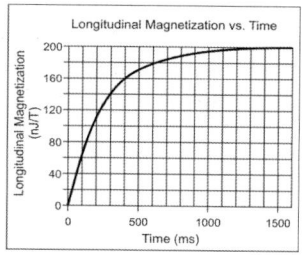

8. In two samples of biological tissue, the application of the rf field sets the net magnetization, M_z, to zero. A) The first sample has a spin–lattice relaxation time of $T_1 = 1.2$ s. How long will it take for M_z to recover to 95% of its equilibrium value? B) The second sample has a spin-spin relaxation time $T_2 = 100$ ms. How long will it take for the transverse magnetization to decay to 37% of its starting value?

Notes

Chapter 1

i. Figure 1.1 was kindly provided by the Österreichische Zentralbibliothek für Physik, University of Vienna.

ii. Dr Nino Čutić, Jägaregatan 204, 226 53 Lund, Sweden kindly sent me high resolution images of a complete set of line spectra,, including the few that appear in Figure 1.6.

iii. Figure A1.1 was provided by Dr Michael Davidson of the Florida State University Research Foundation. Available online at: http://micro.magnet.fsu.edu/

iv. Figure 1.5 is shown in dynamic form in the Flash Animation, *The Bohr Model as seen by a Classical Physicist* at: http://www.physics.utoronto.ca/~key/Flash_Files/Bohr_atom_colla pse.swf

v. My colleague, Dr David Harrison, has developed a large number of excellent Flash animations that provide beautiful visual explanations of many physical processes. There are two animations with relevance to Chapter 1 at http://www.upscale.utoronto.ca/; go to the Flash Animations section, follow: the Quantum Mechanics link to *the Bohr Model*; the Electricity and Magnetism link to *The Electric Field of an Oscillating Charge*, and *Electric and Magnetic Fields of an Oscillating Charge*.

vi. http://www.colorado.edu/physics/2000/index.pl is a wonderful site that describes radiation and modern physics in an interactive and simple way.

Chapter 2

i. Figures 2.2, 2.3, and 2.6 are reproduced with permission from the Institute of Physics, and Oldham (2001).

ii. Figures 2.9, 2.10, 2.11 and 2.12 are adapted with permission by the Institute of Physics, and Michael (2001).

iii. Figure 3.7 and 3.8 is reproduced with permission from the Institute of Physics, and Badawi (2001).

iv. Figure 2.7. and the data in Table A2.2 use, are copied, with permission, from the NIST Web site at http://physics.nist.gov/ PhysRefData/XrayMassCoef/tab4.html.

v. Dr Harrison's Flash animations with relevance to the interactions of X-rays with matter discussed in this chapter can be found at http://www.upscale.utoronto.ca/; follow the link to <u>Nuclear</u>, then *Pair Production*, and *The Interaction of X-rays With Matter*.

vi. The interactive Flash animation, *the Interaction of X-rays traversing matter*, also shows the processes discussed in §2.2: http://www.physics.utoronto.ca/~key/Flash_Files/X-Ray_in_ Matter.swf

Chapter 3

i. The excellent site http://www.colorado.edu/physics/2000/index.p has a variety of resources of relevance to the physics in this chapter. Got to <u>Einstein's legacy</u> and <u>CAT scans</u>.

ii. Dr Harrison's Flash animation with relevance to the processes discussed in this chapter can be found—at http://www.upscale. utoronto.ca/ —by following the link to <u>Nuclear</u>, then 'Nuclear Decays'.

Chapter 4

i. Figure 4.2 and the data in Table A4.1 use, are copied, with permission, from the NIST Web site at http://physics.nist.gov/ PhysRefData/XrayMassCoef/tab4.html.

ii. Dr Dave Rogers kindly sent me the photograph in Figure 4.3, which originally appeared as the cover of Physics in Canada, 58(2), (2002). The Monte Carlo dose engine, developed by Iwan Kawrakow of NRC, is driven by a new electron beam model developed by MDS-Nordion; the rendering was produced by Tomas Lundberg of MDS-Nordion. Reprinted with permission from the Canadian Association of Physicists (See Cover).

Chapter 5

i. Most of the information in this chapter is publicly available from a variety of sources: the Health Physics Society, the International Commission on Radiological Protection, the United Nations Scientific Committee on the Effects of Atomic Radiation, the US Environmental Protection Agency, the US National Council on Radiation Protection and Measurements, and the World Health Organization. The pie charts in Figures 5.1 and 5.2 are drawn from an overview of the data in relevant publications of these organizations, and are intended to provide a representative 'average' visual overview of the contributions to the background radiation.

ii. Dr Ted Rockwell, a nuclear pioneer, gave me permission to use his original graphical compilation of background radiation shown in Figure 5.4, and Dr Jerry Cuttler, President of Cuttler & Associates Inc., kindly sent me a high-resolution copy. Sadly Ted Rockwell died before the publication of this book.

Chapter 6

i. The main reference for this chapter is the excellent and readable article by Stephen Keevil [Keevil, 2001], many of whose figures are reproduced with permission.

ii. Dr. Raymond V. Damadian, founder and President and Chairman of FONAR Corporation, generously sent me the photo that appears in Figure 6.10. The full caption reads: "The first attempt for a human scan was with Dr. Raymond V. Damadian sitting in *Indomitable*, the world's first MR scanner that he and his colleagues built. A blood-pressure cuff was affixed to his right arm, an EKG was wired to this chest, and oxygen was kept handy. The cardiologist (standing at left in the above photo) was there in case the magnetic field produced any strange cardiac effect on Dr. Damadian. No signal was received from the scanner. The team decided that Dr. Damadian was oversized for the cardboard vest housing the antenna and that he must have detuned it. A thinner "guinea pig" was needed." http://web.mit.edu/invent/a-winners/a-damadian.html describes the contributions of Dr. Damadian in text and video format.

iii. The excellent and detailed on-line text of J. Hornak, *the Basics of MRI*, available at http://www.cis.rit.edu/htbooks/mri/ gives a comprehensive treatment beyond the level of this book, with clear animations of spin systems, and a wealth of easily accessible MRI photographs and sounds. Dr Hornak's ideas inform much of section 6.3.1, particularly the design of Figures 6.12 and 6.13.

iv. Dr Harrison has a wonderful Flash animation, a dynamic version of Figure 6.1, at http://www.upscale.utoronto.ca/, *Precession of a Spinning Top*.

v. The Flash animation, *MRI-application of a 90 degree pulse*, available at http://www.physics.utoronto.ca/~key/Flash_Files/MRI_the_Movie.swf is an interactive and dynamic view of the processes described in Figures 6.5 to 6.8.

Worked Examples

These worked problems use a 'standard' multi-step method of approaching problems—model, guess, solve, substitute, check, assess—that has been found to be important in improving students' problem-solving abilities.

1. Number of Atoms in a Given Mass

A gold foil has a mass of 15 g. How many atoms are there in this foil?
Model: We know that a mole of an element contains N_A atoms. So we have to find out how many moles there are in 15g of Gold. Very closely, one mole of an element equals the mass number in g. The mass number of gold is 197. Let N be the number of atoms in the foil.
Guess: Less than N_A; about 1/197 of N_A, so around 10^{21}.
Solve: Let m be the mass of the gold foil, $m = 15g$. 1 mole of Au, containing N_A atoms, has a mass of 197g. Thus there are N_A/A atoms in one gram. So number of atoms in m g is $m\, N_A/A$.
Substitute: No. of atoms in foil,
$$N = m\, N_A/A = 6.02 \times 10^{23} \times 15/197 \text{ (atoms/g)} \times \text{(g)}$$
$$= 4.58 \times 10^{22} \text{ atoms.}$$
Assess: OK!

2. Dose in Air

Given that it takes 33.7 eV to produce one ion pair in air, calculate the dose in air, D_{air} , in Grays, produced by an X-ray exposure of X R. (The answer is given by the formula quoted in §2.3.5).
Model: The dose is the energy deposited in 1 kg (definition). The product of the energy required to produce an ion pair and the number of ion pairs produced by 1 R will yield the total energy deposited by 1 R. We know that 1 R produces 2.58×10^{-4} C of charge in 1 kg of dry air (definition). Each ion pair has a charge equal to the electronic charge, so we can calculate the number of ion pairs.
Guess: Hard to guess. We might remember that 1 R delivers about 1 rad, or about 10mGy $= 10^{-2}$ Gy—so might guess about $10^{-2}\, X$ Gy.

Solve: Let $E(ip) = 33.7$ eV : and $Q = 2.58 \times 10^{-4}$C where $E(ip)$ is the energy of the $N(ip)$ ion pairs. If e is the electronic charge, $N(ip) = Q/e$. Then energy deposited in 1 kg of dry air by X R is :

$$D_{air} = E(ip) \times N(ip) = X \times E(ip) \times Q/e$$

Substitute: Energy deposited by X R in 1 kg of dry air
$$= X \times (33.7 \text{ eV}) \times (2.58 \times 10^{-4}\text{C})/(1.6 \times 10^{-19}\text{C})$$
$$= X \times (5.43 \times 10^{16} \text{ eV}) \times (1.6 \times 10^{-19} \text{J/eV})$$
$$= (8.69 \times 10^{-3} X) \text{ J}$$
Thus $D_{air} = 8.69 \times 10^{-3} X$ J/kg $= 8.69 \times 10^{-3} X$ G, and X R will give a dose in air of $D_{air} = \mathbf{8.69 \times 10^{-3}} X$ Gy
Assess: $8.69 \times 10^{-3} \cong 10^{-2}$.   So OK!

3. Resolution of X-rays

Mammograms are used to search for cancerous tumours in women's breasts. The breast is held steady between two parallel plates and the X-ray taken. Tumours have a density about 120% of healthy tissue. What is the minimum size of tumour located in 7 cm of breast tissue that can be detected by 40 keV X-rays? Assume that X-ray films usually allow contrasts of the order of 2 % (i.e. the ratio of the exposures at the X-ray film of a path including the tumour and a path that traverses only healthy tissue must be less than 98% to be detectable.)

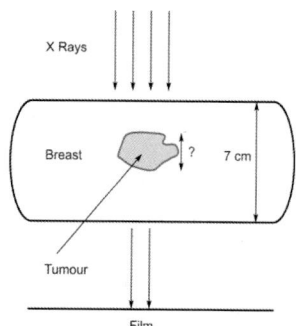

Model: A diagram would be useful here.

Since this question concerns resolution, we use the linear attenuation coefficient to measure the percentage of the original beam that survives its passage through tumour and tissue. Since tumours are denser than tissue, we'd expect that they will attenuate the X-rays more than the surrounding tissue.   The intensity (fluence) of a beam of X-rays decreases exponentially with depth. We have tables for healthy tissue in Chapter 2, but are not given a value for the linear attenuation coefficient for the tumour. However, we know that the mass attenuation coefficient is approximately equal for a variety of tissues at X-ray energies.

Guess: Obviously less than 7 cm! Optimistically we might hope for a few mm. Let's see!

Solve: Use photon fluence as the measure of the radiation reaching the X-ray film. Let d_t be the total depth of healthy tissue and d_c be the depth of the cancerous tumour that we need to find.

Let μ_t, ρ_t be the linear attenuation coefficient and the density of healthy tissue respectively and let μ_c, ρ_c be the linear attenuation coefficient and the density of cancerous tissue respectively. Let $p\%$ be the contrast limit: $p = 98$.

Table A2.2 lists the mass attenuation coefficients for tissue. The linear attenuation coefficients of healthy and cancerous tissue can then be found by multiplying by the density of each:

$$\mu_t = (\mu/\rho)_t \rho_t \text{ and } \mu_c = (\mu/\rho)_c \rho_c$$

Assume that, to a good approximation, the mass attenuation coefficient of healthy tissue equals that of cancerous tissue:

$$(\mu/\rho)_t = (\mu/\rho)_c = 0.239 \text{ cm}^2/\text{g at 40 keV from table A2.2.}$$

The photon fluence emerging from a path of length d_t that includes only healthy tissue is: $\Phi_t(x) = \Phi(0)exp(-\mu d_t)$.

The photon fluence emerging from a path that includes both healthy and cancerous tissue is: $\Phi_{t+c}(x) = \Phi(0)exp\{-\mu_t(d_t - d_c) - \mu_c d_c\}$.

The ratio $\Phi_{t+c}(x) / \Phi_t(x)$ must have a maximum value of p/100:

$$
\begin{aligned}
\Phi_{t+c}(x)/\Phi_t(x) \le p/100 &= exp\{-\mu_t(d_t - d_c) - \mu_c d_c + \mu_t d_t\} \\
&= exp\{\mu_t d_c - \mu_c d_c\} = exp(\mu_t - \mu_c)d_c \\
&= exp\{(\mu_t/\rho_t)\rho_t - (\mu_c/\rho_c)\rho_c\} d_c \\
&= exp\{(\mu_t/\rho_t)(\rho_t - \rho_c)\}d_c
\end{aligned}
$$

using the assumption that $(\mu/\rho)_t = (\mu/\rho)_c = 0.239 \text{ cm}^2/\text{g}$

Substitute: $\{(\mu_t/\rho_t)(\rho_t - \rho_c)\}d_c \ge \ln(100/p)$

Or $\quad \{0.239 \times (1.2 - 1.0)\}d_c \ge \ln(100/p)$,

which yields: $d_c \ge 0.020/(0.239 \times 0.95 \times 0.2) \ge \textbf{4.4 mm}$

(using the value for tissue density given in table 2.1).

Assess: The result is not unreasonable – though in practice, with several exposures, mammograms can reach resolutions of close to 0.1 mm.

4.   Energetics of Radioactivity

$^{92}_{42}$Mo decays via beta minus emission to another nucleus, $^{A}_{Z}$X. The daughter of the beta decay chain is an excited state of $^{A}_{Z}$X , denoted by $^{A}_{Z}$X* in the diagram. $^{A}_{Z}$X* then decays by the emission of a gamma ray of wavelength 6.87 pm as shown in the figure below.

Nuclide	Mass (u)	Half-life (hrs)
Mo	98.907 712	60
X (gr. state)	98.906 254	6
Electron mass = 0.000 549 u		
1 u = 931.494 MeV/c^2		

A) What is the mass number of the nucleus $^{A}_{Z}$X ? B) What is the atomic number of the nucleus $^{A}_{Z}$X ? C) Calculate Eγ , the energy of the gamma ray in MeV. D) What is the relationship between Eγ , E$_0$, and E$_1$? E) Calculate the atomic mass of

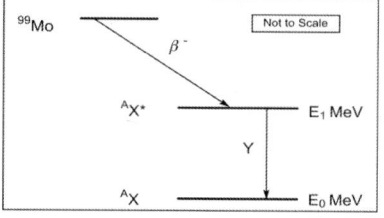

the excited state $^{A}_{Z}$X* in atomic mass units. F) Finally, calculate the maximum energy, in MeV, of the electron emitted in the beta decay of the $^{99}_{42}$Mo to $^{A}_{Z}$X* (neglect any effects of nuclear recoil).

Model and solution: A) The mass number is A and counts the number of protons plus the number of neutrons in the nucleus. Since this number is unaffected by either beta or gamma decay, A= mass number of the $^{92}_{42}$Mo , or **99.**

B) The atomic number is Z which counts the number of protons. Since carge must be conserved, $^{A}_{Z}$X must have one more proton in its nucleus than $^{92}_{42}$Mo since the beta decay carries off one negative electronic charge. The atomic number of the daughter $^{A}_{Z}$X is = 42+1 = **43.**

C) The energy of the gamma ray, of wavelength 6.87 pm is given by:

$$E_\gamma = h\nu = hc/\lambda$$
$$= (6.63 \times 10^{-34} \times 3 \times 10^8)/(6.87 \times 10^{-12} \times 1.6 \times 10^{-19})\text{MeV}$$
$$= \mathbf{0.181MeV}$$

D) The gamma energy is just the difference in the energy of the higher excited level and the lower level to which is decays; $E_\gamma = E_1 - E_0$

E) The gamma ray energy comes from a mass difference of 0.181 MeV/c². This mass difference in amu is $0.181/931.49 = 0.000\ 194$ u. The mass of the excited state is obtained by adding this mass difference to the atomic mass of the ground state of X,:

$$M(X^*) = (98.906\ 254 + 0.000\ 194)u = \mathbf{98.906\ 448u}$$

F) K_{max}, the maximum energy, in MeV, of the electron emitted in the beta decay is calculated from the difference in mass between the Mo and the X* $=M(Mo) - M(X^*) = (98.907\ 712 - 98.906\ 448) = 0.001\ 264$ u. (Note that we use atomic masses here—no need to worry about the use of the nuclear masses or the electron masses, since the latter cancel out in beta minus decay).

Translated into Mev/c², this gives the value of the maximum kinetic energy of the beta decay of:

$$K_{max} = (0.001\ 264 \times 931.494)\,\text{MeV}/c^2 = \mathbf{1.18\,MeV/c^2}$$

5. Generation of Isototopes

A Mo generator, of initial activity 1TBq is milked every 24 hours. What is the activity of the eluted Technetium 18 hours after the second elution? The half life of Mo is 2.5 days and the half-life of Tc is 6 hours.

Model: A diagram is useful here — use Fig 3.6. Immediately after each elution, the Tc activity has to start from zero again. So first calculate the activity of the Mo after two daily elutions, then adapt the formula:

$$R_2(t) = R_1(t)\{\lambda_2/(\lambda_2 - \lambda_1)\}\{1 - exp[-(\lambda_2 - \lambda_1)t]\}$$

Remember to multiply by the branching ratio of 0.88.

Guess: from the diagram, after 2 days, 18 hours, the extrapolated graph gives an answer of around 40 mCi for an initial Mo activity of 100 mCi. So we might guess at an answer around 0.4 TBq.

Solve:Let the parent Mo be milked after a time time T (= two days); its activity is then given by $R_1(T) = R_1(0)exp(-\lambda_1 T)$. Then the Tc activity starting from zero at time T, begins to rise, with time t, according to

$$R_2(t) = R_1(t+T)\{\lambda_2/(\lambda_2 - \lambda_1)\}\{1 - exp[-(\lambda_2 - \lambda_1)t]\}$$
$$= R_1(0)exp(-\lambda_1[t+T])\{\lambda_2/(\lambda_2 - \lambda_1)\}\{1 - exp[-(\lambda_2 - \lambda_1)t]\}$$

Substitute:

Substitute $\lambda_1 = \ln2/T_1$, $\lambda_2 = \ln2/T_2$, $T = 2 \times 24$ hrs,
$T_1 = 2.5 \times 24$ hrs, $T_2 = 6$ hrs, t $= 18$ hrs,
and $R(0) = 1$ Tbq $= 10^{12}$Bq, t $+T = 66$ hrs.
The result for the activity of the eluted Tecnnetium is thus:
$R_2(t) = $ **0.39 TBq**
Assess: OK!

6. Isotopic Dilution

100μCi of tritium, ^{3}H, was given to a person in the form of tritiated water
that was totally absorbed through the gut. After time for equilibriation, a
blood sample was taken that showed an activity of 92 Bq per ml of fluid.
Calculate the total volume of water in the body.
Model: the ratio of radioactive to stable isotopes is the same in the whole
volume of water as it is in a small sample, assuming complete mixing.
Guess: several litres?
Solve: Initial activity $R(0) = 100$μCi $= 3.7 \times 10^6$Bq.,
The volume of the sampl$e = v = 1$ ml $= 10^{-3}$l
The activity of sample $= r(0) = 92$ Bq
The required volume is
$V = vR(0)/r(0) = (3.7 \times 10^6/92) \times 10^{-3}$l $= 40$l
Assess: Perhaps this is a surprisingly large amount! In fact, the
percentage of water in a baby's body is around 80%; it drops to around
60% for adults, so this problem yields a value that is, in fact, a bit low.

7. Equivalent dose, Effective Dose

Suppose the whole body of a hospital worker is
exposed to a 1R X-ray beam as shown in the
figure. A) what, approximately, is the Equivalent
Dose and the Effective Dose?B) Suppose that
shielding removes half of the X-ray beam, as

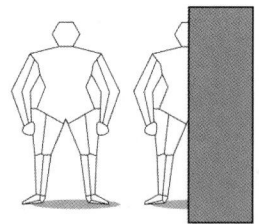

shown in the second figure. What now is the approximate Equivalent Dose and Effective Dose?

Model: Again, use a diagram. The Equivalent Dose is the Absorbed Dose multiplied by the Radiation Weighting Factor, W_R, (=1 for X-ray radiation). The Effective Dose is the Equivalent Dose multiplied by the sum of the Tissue Weighting Factors of the tissues irradiated. Since the whole body is irradiated, this sum, $\Sigma W_T = 1$. Use the 'rule of thumb' estimate: a whole-body exposure of 1R yields an absorbed dose of approximately 1 rad or 10 mGy. For part B), remember that the Equivalent Dose measures the energy deposited per kilogram of biological tissue, so does not change for the half of the body exposed. However, the risk (of contracting cancer, etc.) has obviously decreased by a factor of two.

Guess: Difficult to guess—remembering that the average annual exposure from the environment is 3 mSv, we'd hope this poor person gets not more than several mSv from what appears to be an accident.

Solve:

A) Equivalent and Effective Dose (full body).

Equivalent Dose = 10 mGy × 1 = 10mSv.

Effective Dose = 10 × 1 mSv = 10 mSv.

 B) Equivalent and Effective Dose (half body).

Solve: Equivalent Dose = 10 mSv, Effective Dose = 5 mSv.

Assess: OK!

8.   Biological Excretion

The half life for biological removal of phosphorous from the liver is 18 days. The nuclear half life of ^{32}P is 14.3 days. ^{32}P decays by beta-minus emission, with an average energy of 695 keV, all of which is absorbed in the liver. The liver has a mass of 1.7 kg. A) If there is 10 μCi of ^{32}P in the liver at t=0 what will be the activity in the liver after 10 days? B) What is the dose delivered to the liver in these 10 days?

A):What will be the activity in the liver after 10 days?

Model: Radioactive ^{32}P decays with time due to the natural radioactive decay and normal excretion processes. We need to calculate the effective decay constant.

Guess: The excretion speeds up the rate at which the radioisotope is removed, so we'd expect the effective half-life to be less than 18 days and less than 14.3 days, say 10 days. So about half of the activity might exist in the body after these 10 days, or around 5 µCi.

Solve: $1/T_{eff} = 1/T_n + 1/T_{be}$; So $1/T_{eff} = 1/18 + 1/14.3$.

So T_{eff} = 7.97 days

Activity at time $t = 10$ days is given by

$$R(10) = R(0)exp(-\lambda_{eff}t) = 10 \times exp(-0.693 \times 10/7.97)$$
$$= 4.19\mu Ci$$

Assess: Looks OK.

　　　　B):What is the dose delivered to the liver in these 10 days?

Model: Each nucleus that decays contributes 695 keV of energy. So we find the number of nuclei that have decayed in the body in 10 days, multiply by this energy per decay, and divide by the liver mass to get the absorbed dose.

Guess: As usual, we might expect mGy or even µGy.

Solve: $N(0) = R(0)/\lambda_n$ and $N(10) = R(10)/\lambda_n$

Thus the number of nuclei that have either decayed or been excreted = $N(0) - N(10) = \{R(0) - R(10)\}/\lambda_n$. (Note that from §4.5.3 the ratio of those which decayed before they were excreted to those which were excreted is given by $dN_n(t)/dN_{eff}(t) = \lambda_n/\lambda_{eff}$)

Thus the energy deposited by these nuclei

$e_\beta = (\lambda_n/\lambda_{eff})\{R(0) - R(10)\}/\lambda_n = e_\beta \{R(0) - R(10)\}/\lambda_{eff}$

where $e_\beta = 695$ keV is the energy of each disintegration.

Thus dose, $D_m(10) = e_\beta \{R(0) - R(10)\} \times T_{eff}/(0.693 \times m)$

where $m = 1.7$ kg, the mass of the liver.

Substituting, the result for the dose to the liver is $D_m(10) = $ **14.0 mGy**

Assess: Looks about right.

9. Constant Dose Rate

In some therapeutic treatments it is desirable to inject a constant amount of a radioisotope at regular intervals in order to establish an approximately constant dose rate to the patient over time. If the radioisotope has a decay constant of λ, each injection has an activity of $R(0)$, and T is the time between injections, calculate the activity in the patient's body after a time nT, where n is an integer. (You will need to know how to calculate the sum of a geometric series.). Thus calculate the constant dose rate that will be established after a long time in a patient of mass m, if each decay contributes and energy of e_n.

Model: The dose rate is the rate at which the energy enters the tissue, expressed in Grays per unit time. All we need here is the expression for the decay of the activity with time. Each activity will decay exponentially for time T, when the next injection is delivered; the calculation will be repetitive.

Solve: After a t ime T, the initial activity will have fallen to: $R(T) = R(0)exp(-\lambda T)$. Then a new injection adds $R(0)$, so at time T, the total activity is:

$$R(T) = R(0)exp(-\lambda T) + R(0) = R(0)\{1 + exp(-\lambda T)\}.$$

After another period of time T, this activity will have fallen exponentially, so that: $R(2T) = R(T)exp(-\lambda T) = R(0)\{1 + exp(-\lambda T)\}exp(-\lambda T)$. Then a new injection adds $R(0)$ so at time $2T$, the total activity is:

$$R(2T) = R(0) + R(0)\{1 + exp(-\lambda T)\}exp(-\lambda T)$$
$$= R(0)\{1 + exp(-\lambda T) + exp(-2\lambda T)\}.$$

Repeating this procedure, we find that at time nT, the activity is:
$$R(nT) = R(0)\{1 + exp(-\lambda T) + exp(-2\lambda T) + \ldots\ldots\ldots exp(-n\lambda T)\}$$
This is a geometric series, in which each term is equal to the former term multiplied by $exp(-\lambda T)$, so that the sum is

$$R(nT) = R(0)\{1 - exp(-n\lambda T)\}/\{1 - exp(-\lambda T)\}.$$

After many injections, when n is large, the activity reaches a constant value of $(\infty) = R(0)\{1 - exp(-\lambda T)\}^{-1}$.

The constant dose rate $= e_n R(0)/[m\{1 - exp(-\lambda T)\}]$.

Assess: Since the units of $R(0)$ are per unit time, the units of the dose rate are energy per unit time, divided by mass, or Grays per unit time, as required. As a check on the reasonableness of the answer it is useful to consider what happens at an extreme case; when T is very short, the dose rate becomes very large, as expected.

10. Radiation Risk

What is the increased chance of dying from a chest X-ray (assume the worst scenario!)

Guess: Since it is a fairly common medical procedure, let's hope the risk is small. Guess that it is less than a hundredth of a percent.

Method: §5.1.2 gives estimates for the dose delivered by chest X-ryas, and §5.2.2 gives the LNT estimate of the risk coefficient of 5×10^{-5} times the number of mSv received.

Solve: If a chest X-ray can deliver 250μSv, as §5.1.2 suggests, the risk is $250 \times 10^{-3} \times 5 \times 10^{-5} = 0.00125\%$ over and above the "normal" risk of 25%.

Assess: Much better than we might have expected!

11. Magnetic Resonance Imaging

Explain in detail why MRI procedures are much safer for the patient than X-ray or Nuclear Medicine investigations.

Model: The danger from X- and nuclear radiation arises from ionization in the tissues of the body. The energy required to ionize is typically in the keV or MeV. In MRI the only radiation that enters the body are the photons from the rf radiation. If their energy is small compared to the ionization energy, no damage will be done. Let's calculate!

Solve: The typical (Larmor) frequency is given by $\nu = \gamma B_0 \cong 42 B_0$, with ν in MHz and B_0 in T. Typical MRI magnetic fields are in the region of 1 to 5 T, so typical frequencies are 40 to 200 MHz. At the highest frequency, the energy of each photon is $e_\gamma = h\nu \cong 10^{-6}$eV. Photons of this very low energy will do no damage to the human body!

Bibliography

Burns, D.M. and MacDonald, S.G.G. (1975) *Physics for Biology and Pre-medical Students* (Addison-Wesley).

Dendy, P.P. and Heaton, B. (1987) *Physics for Radiologists* (Blackwell).

Hobbie, Russell K. (1978) *Intermediate Physics for Medicine and Biology* (3rd ed., Springer).

Gale, Robert Peter and Lax, Eric.(2013) *Radiation, what it is, what you need to know* (Alfred A. Knopf, New York).

Johns, H.E. and Cunningham, J.R. (1983) *The Physics of Radiology*, 4th Ed. (Thomas, Springfield, Ill.)

Prince, Jerry L. and Links, Jonathan M. (2006) *Medical Imaging Signals and Systems* (Pearson Education Inc.)

Tuszynski, J.A. and Dixon, J.M. (2002) *Biomedical Applications of Introductory Physics* (John Wiley and Sons, Inc.)

References

Badawi, R.D. (2001). Nuclear Medicine , *Physics Education* 36, pp. 452-459.

Boreham D.R., Dolling, J-A- , Misonoh J., and Mitchel. R.E.J. (2002). Radiation-induced teratogenic effects in fetal mice with varying Trp53 function: Influence of prior heat stress, *Radiation Research* 158(4), pp. 449-457.

Chernobyl Forum, (2003-2005). *Chernobyl's Legacy: Health, Environmental and Socio-Economic Impacts, Second revised version.*

CRC (1982). CRC Handbook of Environmental Radiation. Editor Alfred W. Klement Jr. CRC Press, Boca Raton.

Davidson, Michael, Magnet Lab, Florida State University Research Foundation, Florida State University. Available online at: http://micro.magnet.fsu.edu/.

EPA. (US Environmental Protection Agency). Available online at: http://www.epa.gov/radon/pubs/citguide.html

HBRRG (1980) (High Background Radiation Research Group, China). Health survey in high background radiation areas in China, *Science,* August 1980, pp. 877-880.

Health Physics Society, (2010). Radiation risks in perspective, revised July 2010, *http://hps.org/documents/risk_ps010-2.pdf*, pp. 1-3.

Hubbell, J.H. and Seltzer, S.M. (2004). *Tables of X-Ray Mass Attenuation Coefficients and Mass Energy-Absorption Coefficients* (version 1.4). Available online at: http://physics.nist.gov/PhysRefData/XrayMassCoef /tab4.html

Hornak, Joseph P. (2011). *The Basics of MRI*. (ebook) Available online at: http://www.cis.rit.edu/htbooks/mri/

ICRP(1975). (International Commission on Radiological Protection). Reference Man: Anatomical, Physiological and Metabolic Characteristics. *ICRP Publication 23*. Pergamon Press, Oxford.

ICRP (2007). The 2007 Recommendations of the International Commission on Radiological Protection (Users Edition). *ICRP Publication 103. Ann. ICRP* 37 (2-4).

ICRU (1989). (International Commission on Radiation Units and Measurement), Tissue Substitutes in Radiation Dosimetry and Measurement, *ICRU Report 44 of the International Commission on Radiation Units and Measurements* (Bethesda, MD).

IRSN (2012). (Institute for Radiological Protection and Nuclear Safety). Available online at: http://www.irsn.fr/EN/Research/publications-documentation/radionuclides-sheets/environment/Pages/carbon14-environment.aspx

Jagger, J. (1998). Natural background radiation and cancer death in Rocky Mountain states and Gulf Coast states, *Health Physics*, 75, pp. 428-430.

Jaworowski, Zbigniew (1999). Radiation Risk and Ethics, *Physics Today* 52(9) pp.24-29.

Keevil, Stephen F. (2001). Magnetic resonance imaging in medicine, *Physics Education*, 36, pp. 476-485.

LBNL Tables of Isotopes. Available online at: http://ie.lbl.gov/toi.html

McNeill, D.E.S. and McNeill, K.G. (2002) *Notes for PHY238Y* (Kishmul Resources, unpublished.

Michael, Greg (2001). X-ray Computed Tomography, *Physics Education*, 36, pp. 442-451.

NASA.(National Aeronautics and Space Administration). Available online at:
http://imagine.gsfc.nasa.gov/docs/science/know_11/emspectrum.htm
httpp://imagine.gsfc.nasa.gov/Images/science/EM_spectrum_full.jpg.

NIST. (National Institute of Standards and Technology, Gaithersburg, MD.) Available online at: http://physics.nist.gov/PhysRefData/ XrayMassCoef /tab4.html

NCRP (1988). (The US National Council on Radiation Protection and Measurements). *Radiation Exposure of the U.S. Population from Consumer Products and Miscellaneous Sources*, NCRP Report No. 095.

NCRP (2009) (The US National Council on Radiation Protection and Measurements). *Ionizing Radiation Exposure of the Population of the United States*. NCRP Report No. 160. Available online at: http://www.ncrponline.org/

Oldham, Mark. (2001). Radiation Physics and Applications in Therapeutic Medicine, *Physics Education* 36, pp. 460-467.

OECD (2011). (Organization of Economic Cooperation and Development). OECDiLibrary. Available online at: http://www.oecd-ilibrary.org/

Rockwell, Ted and Cutler, J.(2013). Private communication.

Rogers, D.W.O. (2002). Monte Carlo Techniques in Radiotherapy, *Physics in Canada* 58(2), pp. 63-77.

Ten Hoeve, John E. and Jacobson, Mark Z (2012). Worldwide health effects of the Fukushima Daiichi nuclear accident. *Energy and Environmental Science,* 5, pp.8743-8757.

Thomas, R.G. (1994). The US radium luminisers: A case for a policy of 'below regulatory concern', *J. Radiol. Prot.*, 14, 2, pp. 141-153.

Tsubokura, Masaharu; Gilmour, Stuart; Takahashi, Kyohei; Oikawa, Tomoyoshi and Kanazawa, Yukio (2012). Internal Radiation Exposure After the Fukushima Nuclear Power Plant Disaster, *Journal of American Medical Association,* 308(7), pp. 669-670.

UNSCEAR (1982) (United Nations Scientific Committee on the Effects of Atomic Radiation) *Ionizing radiation: sources and biological effects*. New York: United Nations. Available online at: http://www.unscear.org/ unscear/ en/publications/1982.html

UNSCEAR (2010) (United Nations Scientific Committee on the Effects of Atomic Radiation) *Sources and effects of ionizing radiation.* ISBN 978-92-1-142274-0. New York: United Nations.

Walinder, Gunnar (2000). Has Radiation Protection Become a Health Hazard? *Medical Physics Publishing,* ISBN: 9780944838969, 166 pages.

WHO (2013). (World Health Organization, Geneva).*Health risk assessment from the nuclear accident after the 2011 Great East Japan earthquake and tsunami, based on a preliminary dose estimation,* WHO publications, ISBN: 9789241505130, 172 pages. Available online at: http://www.who.int/ionizing_radiation/pub_meet/fukushima_risk_asses sment_2013/en/index.html.

Index